AEROSPACE ENGINEERING

AEROSPACE ENGINEERING

Laura Sanchez
Vladislav Slavov

Kruger Brentt
Publishers

2025

Kruger Brentt Publishers UK. LTD.
Company Number 9728962

Regd. Office: 68 St Margarets Road, Edgware, Middlesex HA8 9UU

ISBN: 978-1-78715-320-2

For information on all our publications visit our website at http://krugerbrentt.com/

PREFACE

"Aerospace Engineering" takes readers on a journey through the fascinating world of flight and space exploration, offering a comprehensive exploration of the principles, technologies, and applications that define the field of aerospace engineering. This book serves as an essential resource for students, professionals, researchers, and enthusiasts seeking to understand the complexities and advancements driving innovation in aerospace science and technology.

Aerospace engineering encompasses a broad range of disciplines, including aerodynamics, propulsion, structures, materials, avionics, systems integration, and space exploration. From designing and manufacturing aircraft to developing spacecraft for exploration beyond Earth's atmosphere, aerospace engineers play a vital role in shaping the future of air and space transportation, defense, and exploration.

The primary objective of this book is to provide readers with a comprehensive overview of aerospace engineering, covering key topics such as aircraft design and performance, rocket propulsion, orbital mechanics, space mission planning, and advanced materials and technologies. Each chapter offers insights, case studies, and practical applications to help readers grasp the fundamental principles and cutting-edge developments in aerospace engineering.

Contributions from leading experts in the field delve into various aspects of aerospace engineering, including commercial aviation, military aircraft, unmanned aerial vehicles (UAVs), satellite systems, space launch vehicles, and planetary exploration missions. Special emphasis is placed on the interdisciplinary nature of aerospace engineering, highlighting the integration of engineering, physics, mathematics, and computer science in solving complex aerospace challenges.

As humanity ventures further into the cosmos and explores new frontiers of air and space, it is essential to foster innovation, collaboration, and education in aerospace engineering. By inspiring the next generation of aerospace professionals and promoting interdisciplinary research and development, we can unlock new opportunities for exploration, discovery, and technological advancement.

We extend our sincere gratitude to the contributors whose expertise and dedication have enriched this book. It is our hope that "Aerospace Engineering" will serve as a valuable resource and inspiration for students, professionals, and enthusiasts passionate about the limitless possibilities of flight and space exploration.

Laura Sanchez
Vladislav Slavov

CONTENTS

1 Chapter

GREEN ALTERNATIVES TO HYDRAZINE-BASED MONOPROPELLANT AND BIPROPELLANT ROCKET SYSTEMS

1. INTRODUCTION: AN OVERVIEW

Chemical propulsion methods have been widely used in rocket engines for more than 50 years. Hydrazines are the primary chemical propellants preferred for their high performance and dependable history. Most low earth orbit (LEO) satellite propulsion systems utilize monopropellant hydrazine thrusters. The Israeli Offek LEO satellites have a hydrazine propulsion technology. [1, 2, 3]. Figure 1 shows the top plate of the Offek spacecraft with monopropellant hydrazine thrusters, which is the section of the propulsion module facing space. Figure 2 illustrates the propulsion system module and its schematic, detailing the structure and key elements of a standard monopropellant space propulsion system.

Hazards are crucial when working with materials that require energetic reactions for optimal performance. Risks are present in the technology throughout its life cycle, from manufacturing to disposal, including handling, transit, storage, rocket motor firing, and disposal.

"Green propellants" refers to propellants that help minimize potential risks. According to the European Space Agency (ESA), a "green propellant" is a type of propellant that can minimize negative effects on the environment or individuals it may interact with, while yet meeting the necessary performance standards for a mission. The term "reduced hazard propellant (RHP)" is used to refer to propellants where any potential risks have been minimized. Due to increasing safety concerns worldwide, Europe is currently undergoing a legislative procedure that has made the topic of RHP a top priority. This involves decisions made in the European Parliament about the creation of the European Chemical Agency (ECHA) and the implementation of the REACH regulation for Registration, Evaluation, Authorization, and Restriction of Chemicals in Europe in mid-2007 [4, 5].

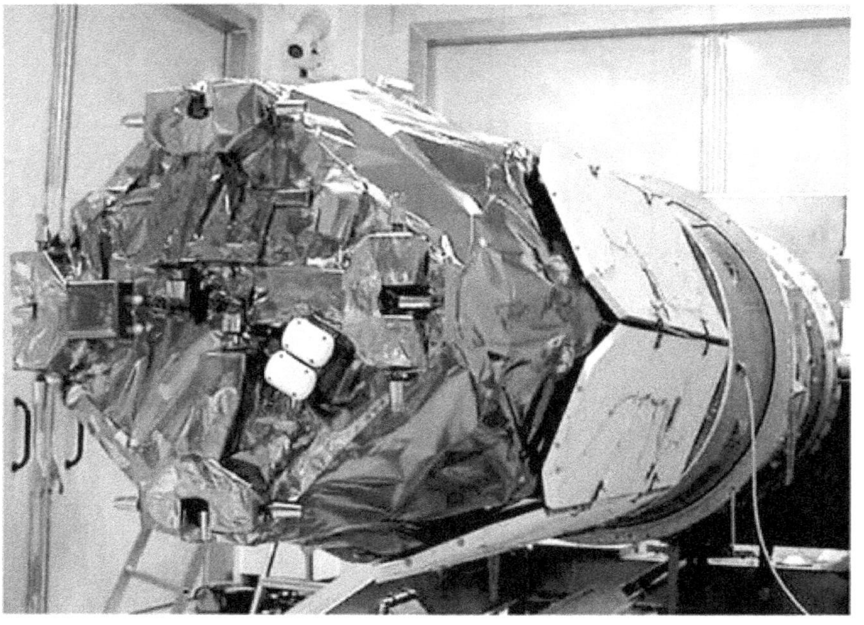

Figure 1. Offek satellite top plate with monopropellant hydrazine thrusters [2].

In the beginning of the era of "green propulsion," alternative propulsion methods to RHP have started to appear. The rate at which these are introduced into space systems is gradual because of the conservative nature of the space propulsion industry. The only environmentally friendly satellite propulsion technology that has been used in space is the ADN-based monopropellant by ECAPS. It was first used in 2010 on the Swedish Prisma satellite and more recently on the American SkySat constellation [7, 8, 9, 10, 11, 12, 13, 14, 15, 16, 17, 18].The ADN-based monopropellant technology has reached the highest technological readiness level (TRL) among the emerging environmentally friendly monopropellants. It is currently being assessed in research and development programs [15] such as the European Horizon 2020,[19, 20, 21, 22, 23],US "green" propulsion evaluation programs, and previously in the European FP7 Green Advanced Space Propulsion (GRASP) program [18]., where the corresponding author also made significant contributions. Monopropellant systems like the ECAPS "green" propulsion system, or RHP, have a history in space applications and are similar to current systems.

Hydrazine-based systems, then for bipropellants, having higher specific impulse Isp and density impulse Isp, the situation is less advanced. Recently, an innovative hypergolic system, based on kerosene and hydrogen peroxide, has been developed by NewRocket© [24], which is similar in performance to MMH/N2O4. The NewRocket Green Propellant (NRGP) hypergolic bipropellant is based on concentrated hydrogen peroxide as oxidizer and on a kerosene-based fuel. NRGP has been made robustly hypergolic by addition of a minute amount of a solid energetic activator to the fuel, which is maintained homogeneously distributed in the fuel by its suitable gelation to a shear-thinning yield-stress fluid. This, while neat HTP and kerosene are not hypergolic. The shear-thinning feature of the fuel enables its full functionality in propulsion systems, including pressurized or pumped feed flow and injection to the reaction chamber, just like any liquid propellant.

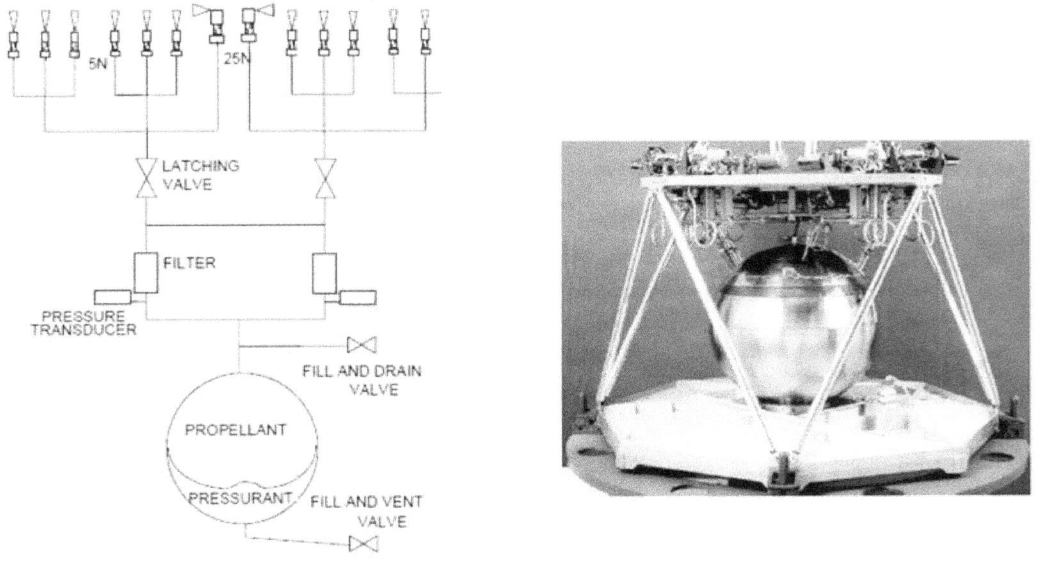

Figure 2. The hydrazine propulsion module [3].

Figure 3 depicts a bipropellant module schematic that is identical to the comparable MMH/N2O4 systems, with the regular components. The thruster assembly consists of a thrust chamber assembly (TCA) with injector, combustion chamber, and converging-diverging nozzle, as well as flow control valve (FCV) that controls fuel and oxidizer feeds. These feeds are provided by regulated pressure from their storage via dedicated manifolds with necessary valves, such as check valves, or inline valves that can be of various types: shape memory alloy actuators (SMA), pyrotechnical valves, or bistable latching valves (LV). The system feed and pressurization are serviced via fill and drain valves (FDV) and fill and vent valve (FVV). Figure 3 also depicts the bipropellant firing test setup which has been realized at the Technion for various NRGP proof-of-concept and development model thruster systems.

Section 2 describes the proposed concept of gradual migration from monopropellant hydrazine propulsion systems to equivalent RHP systems. The concept is based on dual capability of an entire monopropellant chemical space propulsion system. Details are presented of the actual risk reduction program that has been employed for satellite hydrazine propulsion systems that may function also as "green" satellite propulsion systems employing an ADN-based RHP system. Concluding remarks are presented including a conceived way forward with an outlined proof-of-concept firing program [23].

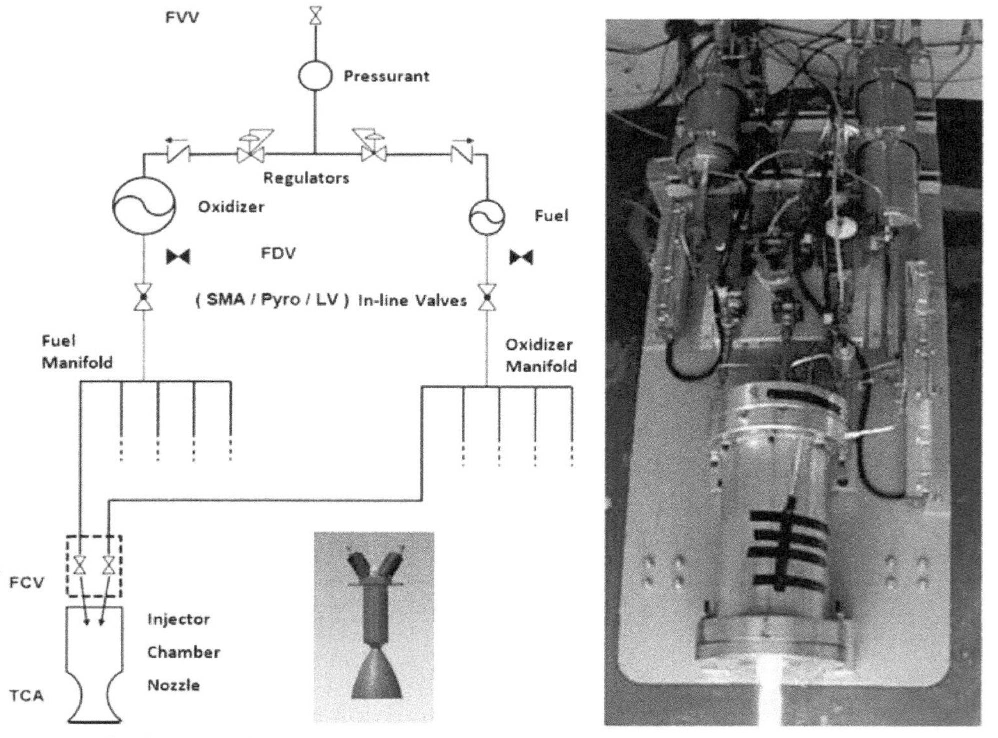

Figure 3.The bipropellant module schematic (left) and test firing setup (right).

Section 3 suggests adopting a similar approach for the hypergolic system utilizing kerosene and hydrogen peroxide, which demonstrates equivalent performance to MMH/N2O4. The results of the fire tests for the proof-of-concept and development model systems, as well as the NRGP fuel rheological characterization, are provided. The results from several engine types show the capacity to use this technology in both pulsating and continuous modes at different levels of thrust. This bipropellant technology provides a promising alternative to current hydrazine-based systems due to its highly reactive fuel and oxidizer, quick ignition, and high characteristic velocity efficiency of over 98%.

2. DUAL CAPABILITY MONOPROPELLANT PROPULSION SYSTEM

2.1. Overview

The idea discussed is a gradual transition to comparable RHP, or "green propulsion" systems. The suggested transition plan involves converting monopropellant systems to RHP by enabling conventional hydrazine systems to function with ADN-based RHP, even shortly before propellant loading. The proposed design is a propulsion system that can make a final decision on fueling with either hydrazine or a RHP. This flexibility will allow the project to advance to a late stage without committing to a specific propellant, facilitating a smooth transition to RHP [23].

The hereby presented concept proposes to go a significant step further than the European "green" Myriade program, which has already sought components compatible both with hydrazine and ADN-based "green" monopropellant, with the notable exception of the thruster assemblies, which will in their case have to be the special ECAPS thrusters [25]. In the "green" Myriade bus, the existing 230 mm propellant tank could be replaced with the one that has a silica-free diaphragm and flown successfully for over 5 years aboard the Swedish Prisma satellite with an ADN-based "green" propellant. For increased propellant capacity, existing larger tanks, using the same materials, can be used. These tanks are also compatible with hydrazine, as has been proven for the diaphragm material for long-term service aboard constellations such as Galileo-IOV and Globalstar-2 that are propelled by hydrazine [26, 27].

The ECAPS dual-mode thrusters and system employ their special thrusters as described in their patent of a thruster and a propulsion system that can be operated either in monopropellant mode or in bipropellant mode [28], as well as their new LMP-103S #1127-3 propellant variant [29]. It combusts at lower temperature giving a specific impulse (Isp), thus comparable with hydrazine, and it is stated that the lower combustion temperature may enable the usage of less expensive materials for the thrust chamber assembly (TCA). This last point is further elaborated in the paragraph below, detailing the risk reduction program of the dual capability monopropellant system, with a concept that has rather drawn some inspiration from the multifuel engine of the Reo trucks, which have had quite a widespread military use [30].

2.2. Risk Reduction Program

This document outlines the initial risk reduction program that has been implemented. The proof of concept demonstrates the dual capability of all propulsion system parts and components, including thrusters, valves, diaphragm tanks, pressure transducers, filters, and pipework. Compatibility of materials and operational suitability have been considered for both hydrazine and RHP in preparation for a comprehensive end-to-end proof by firing testing in a space environment. The program was executed using data analysis and certain tests. Subsequent paragraphs detail various areas where measures have been implemented to mitigate development risks. The primary concern for most components is the compatibility of materials. The propellant must not have a detrimental effect on the components, and the materials used to create the components must not deteriorate the propellant. After that, functional issues are addressed. The chemical reaction that transforms the liquid propellant into high-energy gasses occurs in the thrust chamber assembly (TCA) of the thruster. The catalytic impact on the ADN-based propellant has been demonstrated using the identical catalyst as in the hydrazine thruster. The temperature of the high-energy gasses produced by the basic ADN-based propellant exceeds the heat resistance of the materials used in the hydrazine thruster's TCA. This matter is addressed in a specific paragraph further down. The temperature required to initiate the expected decomposition and oxidation reactions for ADN-based propellant is significantly higher than the 120–180°C needed for hydrazine decomposition. The methods capable of achieving the required higher preheating of ADN-based propellant are outlined in the final paragraph below.

2.3. Compatibility with ADN-Based Liquid Propellants of COTS Hydrazine Systems' Parts, Components, and Materials

The possibility to use commercial off-the-shelf (COTS) construction materials, which are used in typical hydrazine propulsion system components (tubing, valves, filters, etc.), increases the flexibility, improves the reliability, and reduces the costs for introducing the ADN-based reduced hazards' propellants technology on future missions. FOI and ECAPS, who promote the ADN-based RHP FLP-106 and LMP-103S, respectively, have confirmed that most components can be COTS hydrazine propulsion system components [31, 32].

The materials, typically utilized for hydrazine propulsion systems, have been verified to be compatible with ADN-based liquid propellants and specifically with the LMP-103S space-proven aboard the Prisma satellite that was launched in June 2010 and operated successfully in space for 5 years. Table 1 (based on Ref. [33]) presents these for the Prisma satellite.

In the present work, this concept is extended beyond previous works [23], to include also COTS monopropellant hydrazine thrusters, as described below.

2.4. Dual Capability Catalytic Effect with ADN-Based Propellants

The catalytic effect on the ADN-based propellant has been proven with the same catalyst as in the hydrazine thruster. In the TCA, in which the chemical reaction converts the liquid propellant to the necessary high energy gases, there is need for a catalyst to induce such an ignition reaction. In laboratory thermochemical tests, it has been proven that the same iridium catalyst, which decomposes hydrazine, has a definite catalytic effect on the ADN-based propellant.

Differential scanning calorimetry (DSC) analysis has revealed an exothermal peak around 150°C in addition to the endothermic peak around 85°C of ADN melting and the thermal decomposition at 190°C. The peak around 150°C is attributable to the catalytic effect of heated iridium-based catalyst, which does not appear at low temperatures [34]. This has also been found in previously published works [35, 36,

37, 38]. Ignition tests with DSC analysis, such as depicted in Figure 4, demonstrate this effect.

2.5. Temperature and Water Content Effects

The temperature of the high-energy gases in the TCA, using the basic ADN-based propellant, is higher than is normally tolerated by the standard hydrazine thrusters. In order to overcome this limitation, the ADN-based propellants can be adjusted so that the catalytic effect is maintained, whereas the reaction temperature is reduced in order to have the TCA materials of construction within their required temperature limits. For dual use thruster application, this issue can be tackled by

- Using suitable TCA materials with compatibility to higher temperatures and
- Lowering the gas temperature. A combination of both is also possible.

Even though using TCA materials suitable for higher temperatures is not dealt with here, it is notable that these would also be suitable for the lower temperature hydrazine decomposition products, which are the chemically nonoxidizing N2, H2, and NH3.

Table 1. COTS parts and components flown in the Prisma satellite [33].

Component	Supplier	Status
Propellant tank	Rafael	Delta-qual by Rafael
Service valves	Moog	Qualified
Pressure transducer	Bradford	Qualified
System filter	Sofrance	Delta-qual by Sofrance
Latch valve	Moog	Qualified
Thruster	ECAPS	Qualified by ECAPS
Pipes and brackets	ECAPS/SSC	Qualified on STM

As regards lowering the gas temperature, ECAPS presented their new LMP-103S #1127-3 propellant variant that also combusts at a lower temperature, giving a specific impulse comparable with the Isp for hydrazine. They stated that the lower combustion temperature may enable the usage of less expensive materials for the TCA [29]. For their low temperature derivative of LMP-103S, which ECAPS have recently developed, they have conducted hot-firing tests at their facility at FOI-Grindsjön [29, 39], as well as in cooperation with Airbus Safran Launchers (ASL) in the facility at DLR-Lampoldshausen [23, 40, 41, 42, 43]. Although the declared intention of the ECAPS development was to handle significantly lower storage temperatures than specified for traditional storable monopropellants, for example, hydrazine (down to about -30°C), this propellant also exhibited a lower combustion temperature than LMP-103S, giving a specific impulse comparable with the Isp for hydrazine. The low-temperature derivative of the space-qualified LMP-103S was tested in a 22N development thruster, having 20% higher density than hydrazine, combusting at a lower temperature than LMP-103S and with Isp similar to hydrazine [29].

Lowering TCA gas temperatures by the effect of further dilution in water of established ADN-based propellants is expected to be in line with the reduction of the energetic content of the decomposition products. This leads directly to reduction in the temperature of the decomposition gases. The desired effect achieved by this is the possibility to use less demanding materials of construction, but with lower performance. Thruster performance is linked to the temperature directly, as follows. The specific impulse is proportional to the square root of the ratio between temperature and molecular weight of the exhaust gas, or $Isp \sim \sqrt{Tc/M}$ [44]. Therefore, as expected, specific impulse is reduced too, per the square root of temperature.

The specific impulse relation to temperature, $Isp \sim \sqrt{Tc/M}$, is similarly applicable to comparing hydrazine and the ADN-based monopropellant FLP-106 [8]. The specific impulse of FLP-106, as shown in Table 2, is higher than that of hydrazine, in accordance with its considerably higher chamber temperature.

Figure 4. ADN-based liquid monopropellant catalytic decomposition [34].

A detailed investigation and analysis on the influence of the water content on the specific impulse and the thermochemical and density properties of the propellant has been presented in a recent conference by GRASP FP7 group participants. They presented the influence of water content on the ignition process and the spray behavior and the influence on the thermal field inside the combustion. The analysis of the spray behavior in vacuum near conditions was investigated by using different blends. For the analysis of the combustion chamber temperatures, the temperatures and the heat flux inside the combustion chamber in relation to the water content were estimated [45], as well as the impact of the water content and the results for a 500N class thruster.

Table 2. Comparison of the properties of the monopropellants hydrazine and the ADN-based FLP-106 [8].

Monopropellant			Hydrazine	FLP-106
Density		g/cm3	1.0037	1.357
Specific impulse based on mass flow	Isp	s	230	259
Specific impulse based on volume flow	Isp	s g/cm3	231	351
Chamber temperature	Tc	°C	1120	1880
All properties at 25°C, Isp calculated for reaction chamber pressure Pc = 2.0 MPa, ambient pressure Pa = 0.0 MPa, expansion ratio ε = 50.				

The following results were obtained in Ref. [45]. Expansion ratio ε = 50 and chamber pressure Pc = 20 bars were assumed, and the reaction was feasible in the investigated concentrations in water, according to the calculations made. The density reduction still leaves the blend with a considerably higher density than hydrazine, with the corresponding gain in density-specific impulse Isp, decreasing to the lowest value of 280 kg s/L. This is nevertheless approximately 22% higher than the value calculated for hydrazine. Blends bringing Isp down to values similar to those of hydrazine are considered. Density and Isp as a function of temperature for water and FLP-106 at various degrees of water content are presented in Figure 5 (from Ref. [45]).

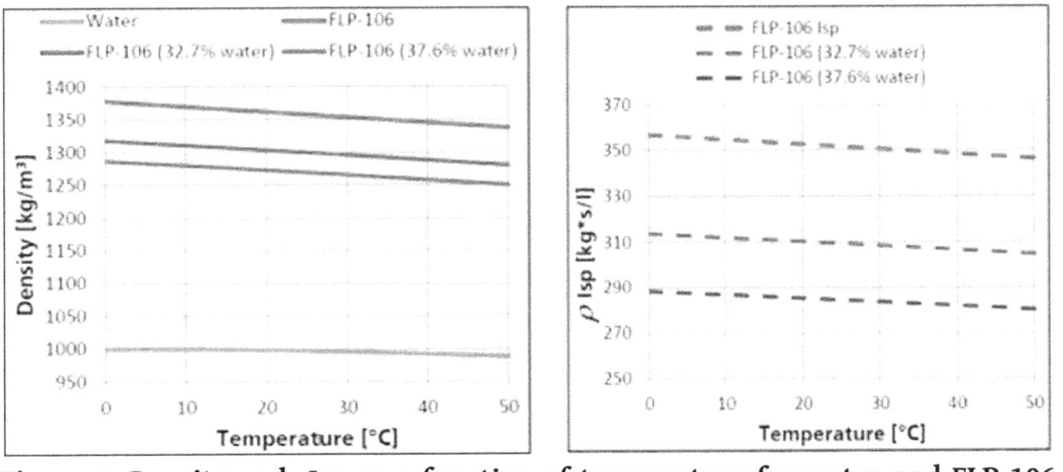

Figure 5. Density and Isp as a function of temperature for water and FLP-106 with various values of water content [45].

2.6. Preheating Temperature Capability

The dedicated ADN-based thrusters are ignited with a preheated catalyst. The ECAPS 1N thrusters, specifically developed for ADN-based monopropellant, use a 10 W heater. The preheating time is 1800 s. In the case of the Prisma thruster, the maximum load during preheating was 9.25 and 8.3 W during firing [46].

The necessary preheat temperature for inducing the reaction of ADN-based propellant is considerably higher than the 120–180°C of hydrazine decomposition. For nominal performance, the required preheat temperatures were in the order of 200–300°C [33]. FLP-106 was experimentally ignited thermally and by resistive heating, within less than 2 ms. an optimal preheating temperature of about 300°C was found where the ignition delay was minimized [47].

Preheating tests have been carried out with a conventional 1N monopropellant hydrazine thruster, with nominal electrical supply voltage in a vacuum chamber that simulates space conditions. These have shown the capability to achieve with a conventional hydrazine thruster the necessary higher preheating of ADN-based propellant, as depicted in Figure 6 [34]. This heating period compares well with the abovementioned 1800 s of the Prisma satellite in-orbit performance.

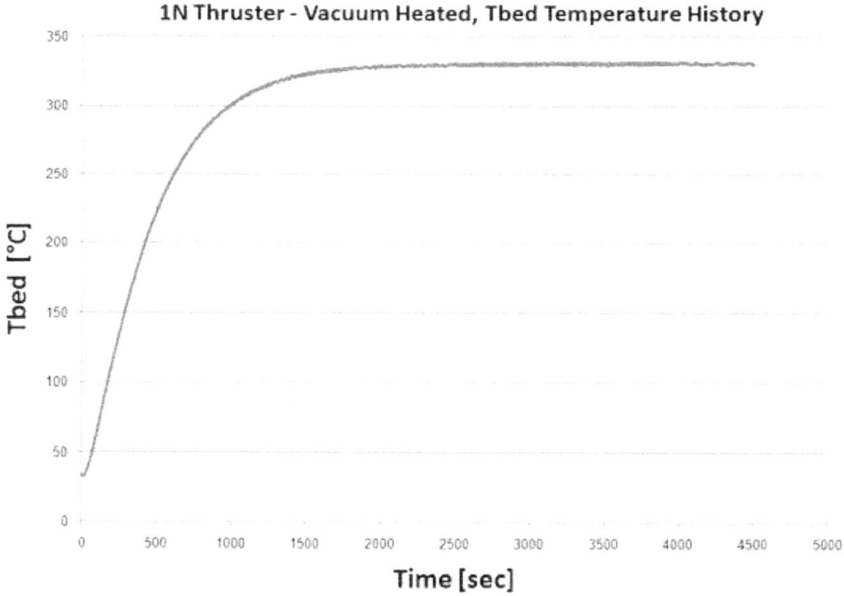

Figure 6. 1N thruster catalyst bed temperature preheat in simulated space conditions [34].

2.7. Proof-of-Concept Firing Program

After removing, within the described initial risk reduction program, the major uncertainties regarding the proposed dual capability of monopropellant hydrazine propulsion systems to operate as equivalent reduced hazards propellant (RHP) systems, a proof-of-concept firing program has been proposed. This program entails end-to-end proof by firing testing in simulated space environment of a representative engineering model (EM) propulsion system.

The proposed test setup is based on existing hydrazine propulsion systems vacuum chamber infrastructure adaptation to ADN-based monopropellant firing, without long-term interference with the capability to continue with hydrazine system tests. The test chamber is the one previously used for the Offek satellite EM firing and depicted in Figure 7 [1].

The difference between the propellants requires attention to aspects of quality and safety to personnel, as well as to those of the testing infrastructure. Primarily, the ADN-based propellant, which is an oxidizer by nature, needs to be very strictly separated from hydrazine, which is fuel by nature. This can be achieved by temporarily disconnecting the existing hydrazine feed lines from the vacuum test chamber setup and maintaining the necessary separation distances according to the materials involved and their quantities. Moreover, the vacuum pump lines should be equally separated, in order to prevent any concern regarding carried over propellants' traces being in contact with each other.

A location independent feed system has been designed for the ADN-based "green" monopropellant, which would be entirely enclosed within the testing vacuum chamber (Figure 7). This is consistent with the end-to-end testing of the entire system, as was done with the hydrazine system EM firing tests.

Here, the handling procedures can be simplified, thanks to the reduced hazards involved with the handling of RHP. The Prisma satellite fueling campaign serves as an example for that, as illustrated in Figure 8. During the launch campaign of the Prisma satellite, the first in-space demonstration of an ADN-based propulsion system, ECAPS loaded the hydrazine and RHP propellants at the Yasny launch base. The handling of ADN-based RHP was evaluated and declared as a "nonhazardous operation" by the Range Safety, so SCAPE suits were not required during the Prisma ADN-based propellant loading operation [20, 48].

The firing program, like any new type of testing, needs to go through the common procedural requirements. These include safety reviews and safety approvals, test procedure preparation and approval, allocation of thruster and propellant, and allocation of the test facility. The procedural requirements have been fulfilled, with the exception of the facility allocation.

Figure 7. Entire system EM testing vacuum chamber [1].

Figure 8. Propellant loading of satellite Prisma [20].

2.8. Concluding Remarks for Monopropellant

An initial risk reduction program has been performed for the concept of dual-capability propulsion systems. This was done by analysis of data as well as by dedicated tests. The program included proof of concept of dual use of all propulsion system parts and components, such as thrusters, valves, diaphragm tanks, pressure transducers, and pipework. The dual use of the propulsion systems' key components, the thrusters, is beyond any previous work. Both material compatibility and actual operation have been justified for both hydrazine and RHP, in view of an eventual system end-to-end proof by firing testing in space simulation environment.

The concept of dual-capability systems may serve as a vehicle toward gradual migration from monopropellant hydrazine propulsion systems to equivalent RHP systems. Hydrazine systems are prevalent in several applications and are still often the systems of choice in space propulsion as well as in other applications. The slow introduction rate of RHP or "green propellants," into space systems due to the conservatism of the space propulsion industry may be expedited thanks to the possibility for gradual conversion by dual capability of conventional hydrazine systems and ADN-based RHP. The presented propulsion system concept may accept last moment decisions on fueling with either hydrazine or with an RHP. This flexibility enables project progress until a very late stage without necessary commitment to either of the propellants, thus allowing a smoother transfer from hydrazine to RHP.

3. COMPARABLE BIPROPELLANT ROCKET PROPULSION SYSTEM

3.1. Overview

This section describes a hypergolic system based on kerosene and hydrogen peroxide, similar in performance to MMH/N2O4 that has been developed by NewRocket© [24]. The NewRocket Green Propellant (NRGP) hypergolic bipropellant is based on concentrated hydrogen peroxide (HTP—high test peroxide) as oxidizer and on a kerosene-based fuel. NRGP is used in a family of bipropellant rocket and gas-generator applications. Neat HTP and kerosene are not hypergolic, while NRGP has been made such by addition of a minute amount of a solid energetic activator to the fuel. The activator is maintained homogeneously distributed in the fuel by its suitable gelation to a shear-thinning yield-stress fluid. Shear-thinning fluids exhibit decreased viscosity with increasing applied shear stresses, such as by pressure gradients (ΔP). The shear-thinning feature of the fuel enables its full functionality in propulsion systems, including pressurized or pumped feed flow and injection to the reaction chamber, just like any liquid propellant.

Usually, decomposition of hydrogen peroxide is achieved using catalyst beds based on silver, platinum, and other materials. Catalyst beds produce high-temperature-decomposed hydrogen peroxide that can burn with a hydrocarbon fuel; however, the system complexity and weight are both increased.

Another method is based on the idea of using catalytic or reactive material (such as metal oxides—MnO2, PbO2, F2O3, etc.) that is dissolved in a liquid fuel. The reactive material decomposes hydrogen peroxide and ignites the fuel, so hypergolic ignition is achieved without the use of a catalyst bed. However, this method requires fuels such as ethanol or methanol that serve as solvents for the reactive material. All these solvents used either alone or with kerosene-based fuels and have relatively low heat of combustion; therefore, the energetic performance of the system is low.

By nature, hydrogen peroxide and kerosene do not ignite upon contact. However, in a gelled fuel, the existence of yield stress assures that particles (reactive or catalytic) can be added without the effect of sedimentation or buoyancy. Gels enable the suspension of reactive or catalyst particles, uniformly distributed in the fuel, without compromising the energetic performance of the system. The use of suspended particles enables a quite large variety of combinations of fuels and oxidizers that can become hypergolic by gelling one of the liquids and adding the proper material.

Natan et al. [49] came up with the idea of embedding reactive particles with hydrogen peroxide in gelled kerosene. Drop-on-drop tests exhibited that this kind of gelled kerosene is hypergolic with hydrogen peroxide as shown in a sequence of photographs in Figure 9. Total ignition delay time was 8 ms. Connell et al. [50, 51, 52] also investigated the issue and showed that it is feasible.

The idea was adopted by a start-up company, NewRocket that proceeded with the development of a prototype motor using gelled kerosene with reactive particles and hydrogen peroxide [24]. Table 3 shows the characteristics of their NRGP propellant in comparison to other candidate propellants.

Here again the specific impulse relation to temperature, $Isp \sim \sqrt{Tc/M}$, is applicable to comparing MMH/N2O4 and the kerosene and hydrogen peroxide-based green bipropellant NRGP. The specific impulse Isp of NRGP, as shown in Table 4, is some 4% lower than that of MMH/N2O4, in accordance with its considerably lower chamber temperature; but it is noteworthy that its Isp is higher by 4% thanks to its 8% higher average density. Experiments have been conducted in a lab-scale motor to verify the feasibility of the idea. The main problems were the atomizers because the particles initially caused plugging of the exit. The problem was solved by changing the type of reactive particles and by increasing the atomizer diameter. The system (Figure 10) was found to operate properly, and by using adequate valves, operation in pulses was achieved as shown in Figure 11.

In the next sections, the stability of the fuel for no phase separation or sedimentation throughout its life cycle is demonstrated by theoretical and experimental considerations.

3.2. Yield Stress Fluids

The NewRocket Green Propellant (NRGP) gelled fuel has been classified as a yield-stress fluid, and this feature has been demonstrated and quantified by tests that included rheological characterization, application of dynamic environment such as acceleration in centrifuge, and real-time storage and handling. The following paragraphs elaborate on that, while being extensively based on Spicer and Gilchrist [53], and are included here in order to make the present chapter quite self-contained.

Yield-stress fluids have the feature of solid-like materials in that they do not flow until a critical stress (σy) is exceeded, after which they flow like a liquid. Modeling such behavior often begins with a nonzero value of the yield stress term σy in the Herschel-Bulkley-Extended (HBE) equation.

$$\sigma = \sigma_y + k\dot{\gamma}^n + \mu_\infty \dot{\gamma} \qquad (1)$$

Here σ is the stress applied on the fluid, $\dot{\gamma}$ is the shear rate, k is a proportionality constant termed the consistency coefficient (viscosity at $\dot{\gamma}$=1), n is a power law exponent termed the flow index, and μ_∞ is the constant viscosity in the very high shear rate range.

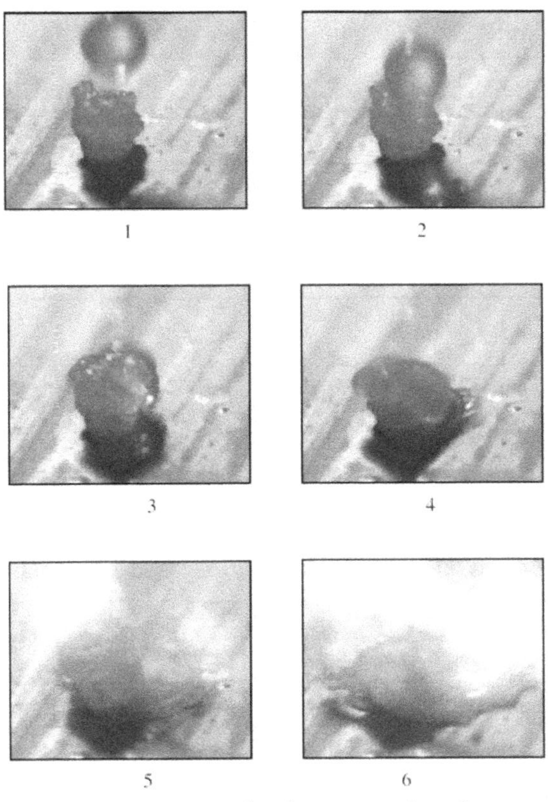

Figure 9. A sequence of high-speed photographs demonstrating hypergolic ignition of hydrogen peroxide with kerosene. The time interval between sequent pictures is 2 ms [24].

Eq. (1) is able to describe power law behavior and includes the additional yield stress term σy. Yield-stress fluids are typically shear thinning and have an exponent of $n < 1$.

The expectation that a fluid might have a yield stress comes from an understanding of the fluid microstructure and its relevant length and time scales. Generally, attractive interactions between colloids, physical crowding of larger particles, and cross-links between polymers or micelles can all provide a finite yield stress to a fluid. Concentration is also a key variable in yield-stress fluids. Very dilute suspensions can have a yield stress but only if the particles attract each other strongly such that they stick together upon collision. The rheology of a suspension gel is highly dependent on whether the particles attract one another strongly enough to form a network that resists flow. Gel microstructure is often a unique function of its processing history because the particle networks can grow, break, and reform under flow.

Table 3. Characteristics of candidate propellants [24].

Propellant	Toxicity	Storability	Cost	Safety	In flight control	Hypergolic ignition
Solid	Low	High	Low	Medium	No	No
Hybrid	Low-none	High	Medium	Medium-high	Yes	No
Hydrazine	High	High	High	Low	Yes	Yes
Ionic propellants ADN-based (HPGP-LMP-103S*)	Low	Medium-high	Medium-high	Medium	Yes	Yes
Bipropellants NTO/MMH*	High	High	High	Medium	Yes	Yes
Electric ion thrusters arcjet	None	High	High	Medium	Yes	–
NRGP	None	High	Low	High	Yes	Yes

Table 4. Comparison of the properties of the bipropellant composition MMH/N2O4 vs. NRGP with kerosene-based fuel/H2O2.

Bipropellant			MMH/N2O4	NRGP
Average density		kg/L	1.2	1.3
Specific impulse based on mass flow	Isp	s	341	328
Specific impulse based on volume flow	Isp	s kg/L	409	426
Chamber temperature	Tc	°C	3125	2580
All properties at 25°C, Isp calculated for reaction chamber pressure Pc = 2.0 MPa, ambient pressure Pa = 0.0 MPa, expansion ratio ε = 50.				

Figure 10. New Rocket lab-scale experimental system.

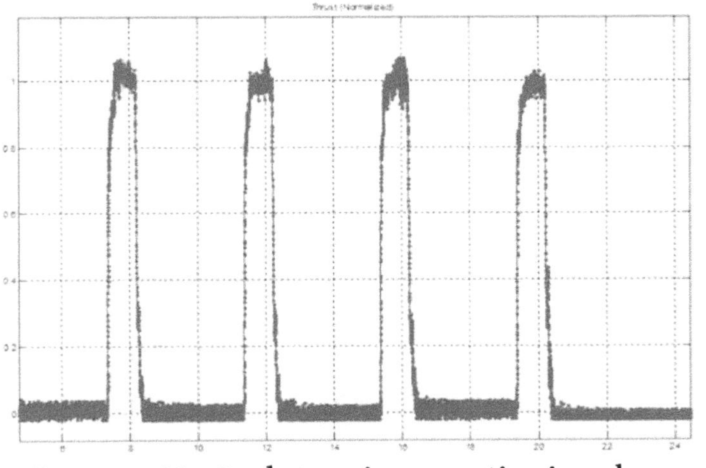

Figure 11. NewRocket engine operation in pulses.

3.3. Yield-Stress Suspension of Particles vs. Stokes Law Settling

The ability of the NRGP fuel, as a yield-stress suspension, to suspend solid particles in the gelled kerosene without any displacement occurring until a shear stress of σy or above is applied, is of key importance. For that, an estimate is made of the magnitude of yield stress required to suspend a given particle. It is important to contrast this treatment with the following Stokes law description of particle settling in a Newtonian fluid, derived via a force balance between the buoyant and drag forces acting on a suspended particle:

$$V = \frac{(\rho_p - \rho_l)\, d^2 g}{18\mu} \quad (2)$$

where the sedimentation velocity at low Reynolds numbers, v, is a function of the particle p and liquid l densities, the particle diameter d, gravitational acceleration g, and the fluid viscosity μ. For other than gravitational accelerations, g would be replaced by the applicable acceleration a.

Rearranging Eq. (2) to solve for viscosity and substituting the height to shelf life ratio for velocity obtains Eq. (3), with sedimentation length and time, h and t, respectively instead of velocity v.

$$\mu = \frac{(\rho_p - \rho_l)\, d^2 g t}{18h} \tag{3}$$

The performance of viscosity with that of a yield stress for the same application can numerically be contrasted by day by day examples [53]. A fluid with a yield stress not exceeded by the acceleration stress of a particle is not described by Eq. (2) because it essentially possesses an infinite viscosity at low stresses and no flow can occur. By taking the ratio of the particle gravitational stress to the fluid yield stress and assuming a hemispherical characteristic area of the yield surface formed, a dimensionless parameter, Y, is obtained to be used to calculate whether a particle will sediment in a yield-stress fluid:

$$Y = \frac{2\pi \left(\frac{d}{2}\right)^2 \sigma_y}{\frac{4}{3}\pi \left(\frac{d}{2}\right)^3 (\rho_p - \rho_l) g} = \frac{3\sigma_y}{d(\rho_p - \rho_l) g} \tag{4}$$

where d is the particle radius and σy is the fluid yield stress. It is worthwhile noting that the critical Y, Ycrit, bounding the states of suspension and sedimentation, is less than unity because of the finite fluid volume yielded by the particle. This means that the yield stress required to suspend a given particle is actually less than the gravitational stress the particle exerts. Simulations give a value of Ycrit = 0.14 [54], while experiments produce Ycrit values between 0.1 and 0.6 [55]. Since the critical criterion can vary significantly, so can the suspension efficiency of a yield-stress fluid. Eq. (4) can be used to estimate the yield stress required to stably suspend a small solid particle by assuming a worst case of a Ycrit = 1. If the worst case application is not satisfying the requirements, then Eq. (4) may be used to remove the extraconservatism by using it as a non-dimensional index. This can be experimentally determined for a specific fluid-particle system using a test in which the suspension stability of a range of particle sizes or densities is recorded for a specified yield-stress fluid and the transition from stability to sedimentation is recorded. The approach described above applies to sedimentation of a dilute suspension of particles through a homogeneous yield stress fluid or, equivalently, of a much larger single particle through a homogeneous suspension of small particles.

It can be demonstrated [53] that yield stress can be a very efficient means of stabilizing particle suspensions because it can entirely prevent any particle motion, whereas viscosity merely slows particle motion.

3.4. Calculation of the Sedimentation Threshold Acceleration of a Particle through the NRGP Gelled Fuel

Rearranging Eq. (4) and replacing the gravitational acceleration g with the applicable acceleration a in order to solve it, and substituting the height to shelf life ratio for velocity, obtains Eq. (5).

$$a = \frac{3\sigma_y}{Y \cdot d(\rho_p - \rho_l)} \quad (5)$$

Using Eq. (5) for a particle with diameter d = 2 μm and density of p = 1.45 g/cc, immersed in a gel with density l = 0.8 g/cc and with a σy of 10 Pa (a conservative order of magnitude representative of the 16 Pa measured in the paragraph below), while assuming a worst case of a Ycrit = 1, the solid particle will start to move when the acceleration reaches a threshold value of a = 23,077 m/s2 or a = 2352 g.

3.5. Calculation of the Sedimentation Velocity and Distance of a Particle through the NRGP Gelled Fuel

When the yield stress σy = 10 Pa, which is the conservative threshold for movement, has been surpassed by applying acceleration 2400 g (in excess of 2352 g), the sediment movement velocity of a particle through fluids can be calculated using Eq. (2). Here the acceleration of gravity g would be replaced by the applicable acceleration a.

$$V = \frac{(\rho_p - \rho_l)\, d^2 a}{18\mu} = \frac{650\frac{kg}{m^3} \cdot \left(2 \cdot 10^{-6} m\right)^2 \cdot (2400 \cdot 9.81) m/s^2}{18 \cdot 100\, Pa \cdot s} = 3.5 \cdot 10^{-9} m/s \quad (6)$$

This, with a gel viscosity measured in experiments of μg=100Pa·s. However, the viscosity derived from the Herschel-Bulkley rheological model coefficients in the following paragraph would be taken into account here.

In 10 s under acceleration of 2400 g, the resultant sedimentation distance is 3.5 × 10-8 m, namely one-thirtieth of a micron sedimentation. In 10 years (10 × 3600 × 24 × 365 = 3.15 × 108 s), this represents sedimentation of 1 m, namely full sedimentation. Therefore, it is important to remember that the extremely high acceleration value, in the order of km/s2, was assumed here just in order to compare to the actually high yield stress of the fuel, while in reality any considerable accelerations are applied on the fuel for very short periods only, such as experienced by space launch.

For a fuel with similar viscosity, but without a yield stress (not being the case here), for gravitational acceleration of 9.81 m/s2, in 10 years the sedimentation would merely be 4 mm.

It can be seen that the sedimentation distance of a particle is proportional to its squared diameter, gravity, and particle density and inversely proportional to the viscosity of the gel. Thus, it is possible to reduce the sediment distance by the following ways: reducing particle diameter, reducing particle density, and increasing the viscosity of the gel.

Based on Technion experience, it can be stated that after storage of a couple of years, there is no degradation in terms of phase separation, sedimentation, agglomeration, ignition delays, etc. For quantitative evaluation of these behaviors, both real-time and accelerated tests are relevant. These were obtained by centrifuge tests for assessing the stability of the gel in accelerations.

To simulate the mechanical environmental loads during a typical rocket launch, which might cause concern regarding gel separation in the tank and propellant feed system, centrifuge tests have been conducted. Example of result obtained for a gelled fuel with a suspended particle with a diameter of 250 μm is depicted in Figure 12, which shows the gel stability as a function of operating time or degree of acceleration. In this experiment, a gelled fuel sample within a test tube was tested in a centrifuge for assessing the influence of two different conditions: firstly, applying constant acceleration (40 g) while varying the time duration (Figure 12 left) on the test and secondly, applying constant duration time (2 minutes) while varying the magnitude of the acceleration (Figure 12 middle). After each centrifuge test, the separated liquid due to the acceleration has been sought in order to be compared with the initial mass to quantify the stability of the investigated gel.

It is important to note that from a visual examination of all samples, no particle sedimentation was observed.

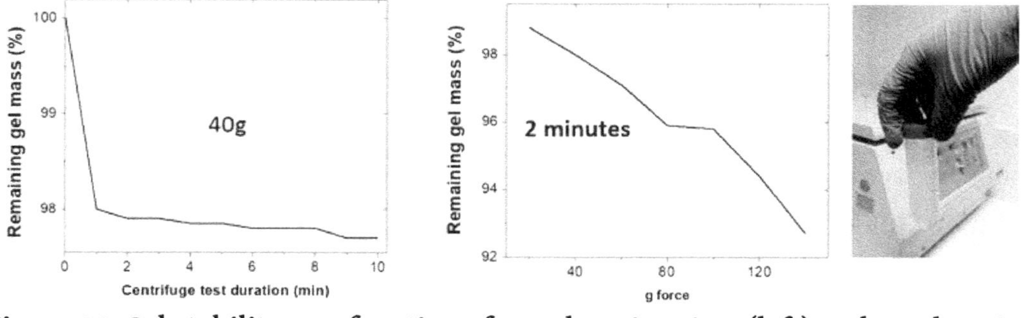

Figure 12. Gel stability as a function of test duration time (left) and acceleration value (middle). On the right is a test tube before centrifuge test.

3.6. NRGP Fuel Rheological Characterization

3.6.1. Measuring System

For characterization of the rheological behavior of the gelled fuels, a TA Instruments AR 2000 rotational rheometer [56] operated in controlled rate mode is being used. The rotational rheometer imposes strain to the liquid and measures the resulting stress for shear rates up to 1000 1/s. Most common test geometries for rotational rheometers are the parallel plates (see Figure 13 right) and the cone and plate. The parallel plates configuration has been used here for gel characterization. A Peltier plate-type temperature regulation system inside the equipment ensures the prescribed controlled fluid temperatures during rheological measurements.

Figure 13. AR 2000 rheometer and parallel plates [56].

3.6.2. NRGP Measured Results

For measuring shear-thinning and thixotropic characteristics, fuel gel samples were subjected to hysteresis loop tests starting with increasing shear rate $\dot{\gamma}$ from 1 to 100 s-1 (up curve) and then reducing back from 100 to 1 s-1 (down curve) while measuring the viscosity μ. In Figure 14, it can be seen that the shear-thinning behavior is significant: μ decreases almost two orders of magnitude in the presented shear rate range. At high shear rates $\dot{\gamma}$ > 102 s-1 (not in the figure), the shear-thinning behavior diminishes due to the destruction of the gel structure and leads to a constant viscosity value μ_∞, which is called upper Newtonian plateau and its value is near the viscosity of the Newtonian neat fluid fuel.

This result is in accordance with the requirements for the rocket engine propellant feeding system, for which the gel fluid passes through a pipe and finally is injected into the combustion chamber. The injectors are small in both length and cross-sectional area, and the fluid remains there for a very short time. The shear rates developed in the injectors due to the sudden decrease in the cross-sectional area are very large, and the shear-thinning effect is dominating.

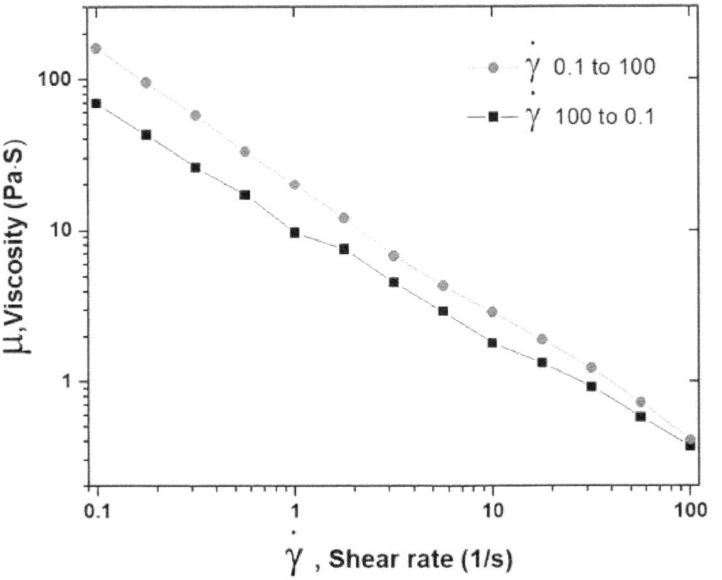

Figure 14. Gelled fuel viscosity as a function of shear rate.

3.6.3. Yield Stress Measurement

The yield stress value of the gelled fuel can be measured by two main methods:

Shearing the fluid at a low and constant shear rate and measuring the shear stress as a function of time. In this method, the yield stress is defined as the maximal measured value as shown in Figure 15a.

Measuring the shear stress vs. shear rate and extrapolating to zero shear rate (Figure 15b) using the Herschel-Bulkley equation, as detailed above for Eq. (1).

Here method (b) was used for measuring yield stress on the rotational rheometer, that is, by extending the flow curve at low shear rates and taking the shear stress y-axis intercept as the yield value. Using this method for an NRGP fuel sample is shown in Figure 16. In this example, the measured yield stress value is 16 Pa.

The three aspects of the shear relevant rheological behaviors of a gel, that is, shear-thinning, upper Newtonian plateau, and yield stress, can be described by a constitutive equation, which is the Herschel-Bulkley extended (HBE) equation, Eq. (1), expressed here in terms of viscosity. This equation describes the dependence of the shear viscosity on the shear rate.

$$\mu = \frac{\tau_0}{\dot{\gamma}} + K \cdot \dot{\gamma}^{n-1} + \mu_\infty \quad (7)$$

As mentioned above, n is the flow behavior index that varies from "0" for very shear thinning materials to "1" for Newtonian materials. A smaller value of n means a greater degree of shear thinning. An example using HBE method for assessing the shear thinning behavior is shown in Figure 17.

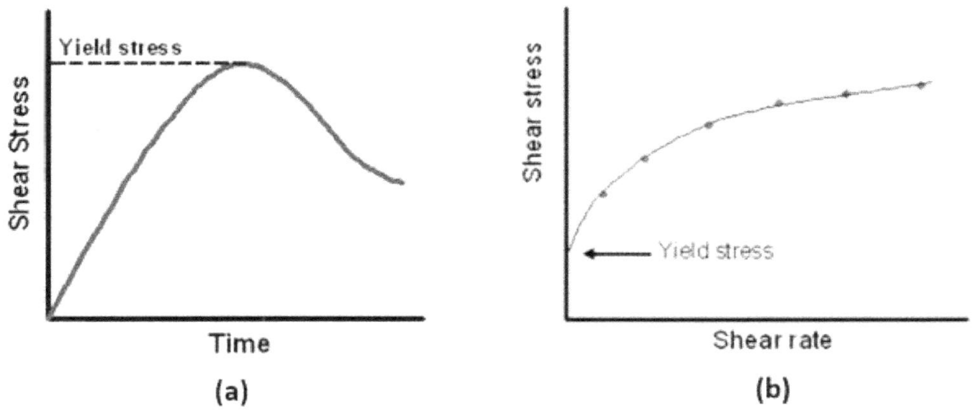

Figure 15. Common methods for measuring yield stress [56]. (a) Shearing a fluid at a low and constant shear rate and measuring the shear stress as a function of time and (b) extending the shear stress vs. shear rate curve at low shear rates.

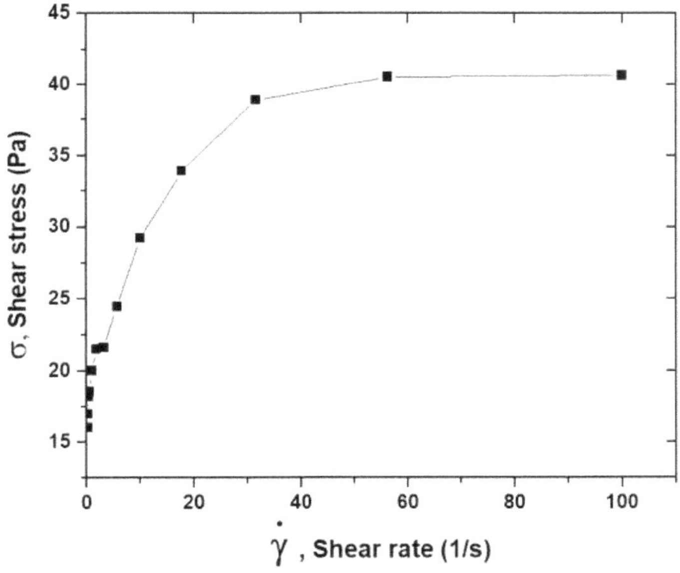

Figure 16. Yield test for an NRGP fuel.

3.6.4. Temperature Effect

The gel fuel temperature has an effect on its rheological properties (shear thinning, yield stress, thixotropic behavior). In general, the yield stress of the gel decreases with increasing temperature since the cohesion forces between the gel molecules are decreasing and their mobility is increasing, thereby the resistance of the gel to deformation is reduced as the temperature increases. Shear thinning and thixotropic behaviors are becoming less prominent at higher temperatures.

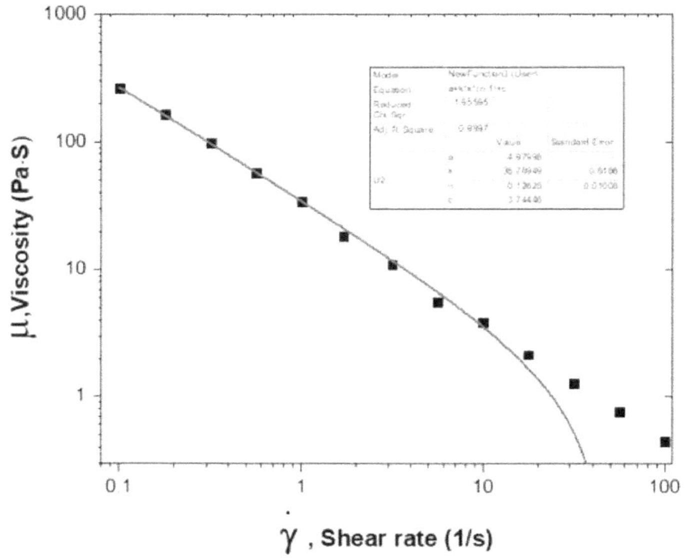

Figure 17. Curve fitting using Herschel-Bulkley extended model.

An example of measurement made for an NRGP fuel gel sample at three different temperatures, -10, +40, and +70°C is shown in Figure 18, which shows that the influence of the temperature on the viscosity values is more prominent at lower shear rates and the shear thinning behavior becomes less prominent as the temperature increases.

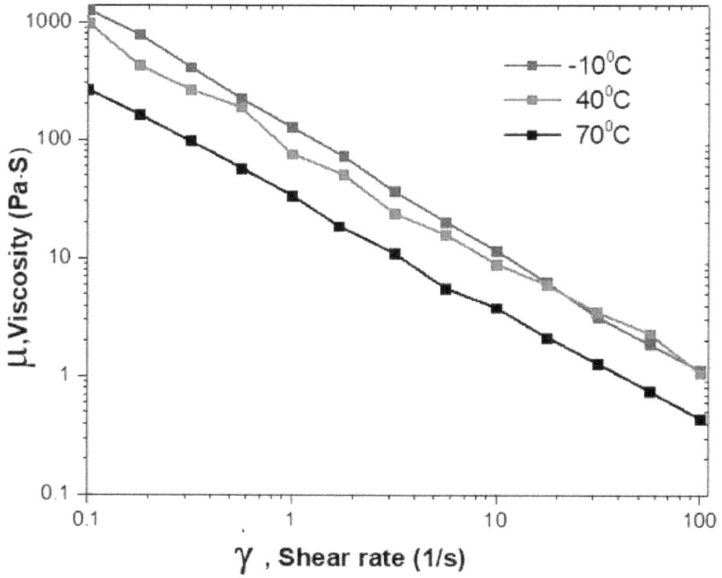

Figure 18.Gelled fuel viscosity as a function of shear rate at -10, 40, and 70°C.

It is important to note that for an expected operational range of -10 to +40°C, the temperature effect is very small.

Visual inspection of a sample that has been exposed to a temperature of 80°C revealed no phase separation or sedimentation at all. Other changes (color, bubbles, pitting) were not observed either.

4. CONCISE SUMMARY

Advanced propellants that are less hazardous than hydrazines have been found as suitable substitutes for traditional propulsion systems. The solutions given here provide continuity in technical idea and reduce costs for introducing new technology to the conservative sector of space propulsion. A design has been created for a dual system using catalytically activated hydrazine monopropellant. The decision on whether to utilize traditional or environmentally friendly "green" propellant can be postponed. Therefore, a seamless technical transfer is facilitated without the need for expensive specialized components.A system development has been introduced for hypergolically activated bipropellants, replacing the hazardous MMH and N2O4 with kerosene-based fuel and hydrogen peroxide oxidizer. Both are non-toxic and commonly used at a low cost in everyday life.

5. AT A GLANCE

Proposals are introduced for environmentally friendly alternatives to monopropellant hydrazine propulsion systems and hypergolic bipropellant systems with comparable performance levels. As we enter the era of "green propulsion," environmentally friendly alternatives to hydrazine propulsion are becoming more prevalent. The integration of these systems into space is hindered by the conservative nature of the space propulsion business, resulting in a delayed rate of adoption. This concept proposes a transition to environmentally friendly propellants by combining traditional hydrazine systems with ADN-based systems. A preliminary risk reduction program has been implemented to actualize the notion. It contains evidence of the feasibility of utilizing all propulsion system components for two different purposes. Demonstrations have shown that the materials are compatible and operational. We introduce a new environmentally friendly hypergolic system for bipropellants using kerosene and peroxide, which offers performance comparable to MMH/N2O4. The proof-of-concept and development model systems' results are displayed. The experimental findings of different engine varieties show the capacity to function in both pulsating and continuous modes, as well as the aptitude to generate varying degrees of thrust. The fuel and oxidizer exhibit strong hypergolic behavior with short ignition delay times, and a distinctive velocity efficiency of over 98%.

REFERENCES

1. Hasan D, Adler S, Oren A, Miller N. Propulsion systems for small satellites. Space Technology. 1995;15(6):375-381
2. Inbar T. 2008. Available from: http://www.iai.co.il/sip_storage/files/5/36055.wmv to https://he.wikipedia.org/wiki/%D7%A7%D7%95%D7%91%D7%A5:Ofek_7_%2B_launcher.jpg [Accessed: August 29, 2008]
3. Hasan D, Jaeger M, Oren A, Adler S, Miller N, Zemer E. Application of satellite hydrazine propulsion system in-orbit monitoring model. In: Proc. 4th Int'l Spacecraft Propulsion Conference, ESA SP-555; 2-9 June 2004; Chia Laguna (Cagliari), Sardinia, Italy. 2004
4. The European Parliament. Registration, Evaluation, Authorisation and Restriction of Chemicals (REACH), Regulation (EC) No 1907/2006; 2006
5. European Chemicals Agency. Agreement of the Member State Committee on the Identification of Hydrazine as a Substance of Very High Concern; 2011
6. Gotzig U. Challenges and economic benefits of green propellants for satellite propulsion. In: EUCASS-2017, The 7th European Conference for Aeronautics and Space Sciences; 3-6 July 2017; Milan, Italy. 2017
7. Anflo K. Concluding a 5 year in-space demonstration of an ADN-based propulsion system on PRISMA. In: Space Propulsion 2016; 2-6 May 2016; Rome, Italy. 2016
8. Larsson A, Wingborg N. Green propellants based on ammonium dinitramide (ADN). In: Hall J, editor. Advances in Spacecraft Technologies. 2011. Available from: http://cdn.intechweb.org/pdfs/13473.pdf
9. Wingborg N, de Flon J. Characterization of the ADN-based liquid monopropellant FLP-106. In: Space Propulsion 2010 Conference; 4-8 May 2010; San Sebastian, Spain. 2010
10. Thormählen P, Anflo K, Sjöberg P, Skifs H. A liquid ADN-based monopropellant for space propulsion. In: Space Propulsion 2010; San Sebastian. 2010
11. Ek S. Qualification of LMP-103S—An ADN-based satellite propellant. In: Space Propulsion 2016; 2-6 May 2016; Rome, Italy. 2016
12. Thormählen P. Space qualification of monopropellant LMP-103S. In: Space Propulsion 2016; 2-6 May 2016; Rome. Italy. 2016
13. Anflo K. High performance green propulsion-on the way for three launches from three continents. In: Space Propulsion 2016; 2-6 May 2016; Rome. Italy. 2016

14. Persson M. HPGP® a flight proven technology selected for multiple LEO missions. In: Space Propulsion 2016; 2-6 May 2016; Rome, Italy. 2016

15. Mulkey H, Miller JT, Bacha C. Green propellant loading demonstration at U.S. range, AIAA 2016-4575. In: 52nd AIAA/SAE/ASME Joint Propulsion Conference. Salt Lake City, UT; 2016

16. Friedhoff P, Hawkins A, Carrico J, Dyer J, Anflo K. On-orbit operation and performance of ammonium dinitramide (ADN) based high performance green propulsion (HPGP) systems, AIAA 2017-4673. In: 53rd AIAA/SAE/ASME Joint Propulsion Conference. Atlanta, GA; 2017

17. Dinardi A, Anflo K, Friedhoff P, On-Orbit Commissioning of High Performance Green Propulsion (HPGP) in the SkySat Constellation, Paper SSC17-X-04; Salt Lake City: Small Satellite Conference; 2017

18. Wurdak M, Strauss F, Werling L, Ciezki H, Greuel D, Lechler R, et al. Determination of fluid properties of the green propellant FLP-106 and related material and component testing with regard to applications in space missions. In: Space Propulsion 2012; Bordeaux. 2012

19. Negri M, Hendrich C, Wilhelm M, Freudenmann D, Ciezki HK, Gediminas L, et al. Ignition methods for ADN-based liquid monopropellants. In: The EU H2020, New Energetics Workshop; 17-19 May 2016; Sweden: FOI Kista. 2016

20. Negri M. Replacement of hydrazine: Overview and first results of the H2020 project Rheform. In: 6th European Conference for Aeronautics and Space Sciences (EUCASS). 2015

21. Negri M, Wilhelm M, Hendrich C, Wingborg N, Gediminas L, Adelöw L, et al. Technology development for ADN-based green monopropellant thrusters—An overview of the RHEFORM project. In: EUCASS-2017, The 7th European Conference for Aeronautics and Space Sciences; 3-6 July 2017; Milan, Italy. 2017

22. Wilhelm M, Negri M, Hendrich C, Wingborg N, Gediminas L, Adelö L, et al. The RHEFORM project-developments for ADN-based liquid monopropellant thrusters, AIAA 2017- 4672. In: 53rd AIAA/SAE/ASME Joint Propulsion Conference. Atlanta, GA; 2017

23. Hasan D, Grinstein D. Dual capability monopropellant rocket propulsion concept. In: Proceedings of the 58th IACAS-Israel Annual Conference on Aerospace Sciences. 2018. pp. 520-533

24. Natan B. Advances in gel propulsion, plenary paper ISICP-PL4. In: 11th International Symposium on Special Topics in Chemical Propulsion & Energetic Materials (11-ISICP); September 2018; Stuttgart, Germany. 2018

25. Lange M, Lein A, Gotzig U, Ziegler T, Anthoine S, Persson M, et al. Introduction of a high-performance ADN based monopropellant thruster on the Myriade propulsion subsystem—Technical and operational concept and impacts, recent advances in space technologies (RAST). In: 6th International Conference on Recent Advances in Space Technologies; 12-14 June 2013; Istanbul, Turkey. 2013

26. Hasan D, Tavor M, Naveh N, Pavlovsky S, Shechter Y, Oren A, et al. Diaphragm tank based hydrazine systems for long space missions. In: Space Propulsion 2010; 3-6 May 2010; San Sebastian. 2010

27. Hasan D, Naveh N, Pavlovsky S, Tavor M, Fiksman T, Zaberchik M. Materials compatibility considerations of satellite propellant tanks with EPN-40 diaphragms. In: Space Propulsion 2008, 5-9 May 2008; Heraklion, Crete, Greece. 2008

28. Bergman G, Anflo K, Dual mode chemical rocket engine and dual mode propulsion system comprising the rocket engine. US Patent Application: 20160115906; 2016

29. Anflo K, Thormählen P, Persson M. Hot-firing tests using a low temperature derivative of LMP-103S, paper a625. In: EUCASS-2013: 5th European Conf. For Aeronautics and Space Sciences; 1-4 July 2013; Munich, Germany. 2013

30. Doyle D. Standard Catalog of U.S. Military Vehicles. 2nd ed. Iola, WI: Krause Publications; 2011

31. Wingborg N, Johansson M, Bodin L. ADN-based liquid monopropellants (presentation). In: ESA 3rd International Conference on Green Propellants for Space Propulsion; 17-20 September 2006; Poitiers, France. 2006

32. Beckel S, Dinardi A. High-performance green propulsion (HPGP) for improved performance, responsiveness and reduced lifecycle cost. In: Space Tech Expo, Satellite & Space Summit; 22 May 2013; Long Beach, CA. 2013

33. Anflo K, Bergman G, Hasanof T, Kuzavas L, Thormählen P, Åstrand B. Flight demonstration of new thruster and green propellant technology on the Prisma satellite, paper SSC07-X-2. In: 21st Annual AIAA/USU Conference on Small Satellites; 13-16 August 2007; Logan: Utah State University. 2007

34. Hasan D, Yamin S, Reiner D, Gottlieb L, Blummer-Ganon B, Dreerman E, Grinstein D. Report on State of the Art in Alternative (Reduced Hazard) Liquid Propellants for Space Propulsion Rocket Engines, RafDocs#1504523; January 2015; 2015. Unclassified

35. Amariei D, Courthéoux L, Rossignol S, Batonneau Y, Kappenstein C, Ford M, et al. Influence of fuel on thermal and catalytic decompositions of ionic liquid

monopropellants. AIAA Paper 2005-3980. In: 41st AIAA/ASME/ SAE/ASME Joint Propulsion Conference & Exhibit. 2005

36. Scharlemann C. Green advanced space propulsion—A project status, paper AIAA-2011-5630-890. In: 47th AIAA/ASME/SAE/ASEE Joint Propulsion Conference & Exhibit; San Diego, California. 2011

37. Robinson JW. Green mono propulsion activities at MSFC, SP2014 paper 2969495. In: Space Propulsion 2014; 18-22 May 2014; Köln, Germany. 2014

38. Robinson J, Beckel S. Green propulsion auxiliary power unit demonstration at MSFC, SP2014 paper 2925788. In: Space Propulsion 2014; 18-22 May 2014; Köln, Germany. 2014

39. Thormählen P and Anflo K, Low-temperature operational and storable ammonium dinitramide based liquid. Patent WO2012166046 A2 & A3; 2012

40. Gotzig U, Telitschkin D, Haecker F, Duelger E. Overview and actual development status of astriums monopropellant thruster family, SP2014 paper 2968006. In: Space Propulsion 2014; 18-22 May 2014; Köln, Germany. 2014

41. Gotzig U, Ziegler T, Fiot D, Wynn J, Persson M, Anflo K. Astriums alternative propellant activities, SP2014 paper 2968009. In: Space Propulsion 2014; 18-22 May 2014; Köln, Germany. 2014

42. Welberg D, Lange M, Anflo K, Persson M, Bahu J-M. Conceptual study of a HPGP reaction control system for a launcher upper stage, SP2014 paper 2965218. In: Space Propulsion 2014; 18-22 May 2014; Köln, Germany. 2014

43. Persson M, Welberg D, Lange M, Bahu J-M, Anflo K. 200N HPGP thruster demonstration for launcher upper stage reaction control system, SP2014 paper 2968097. In: Space Propulsion 2014; 18-22 May 2014; Köln, Germany. 2014

44. Klapötke TM. Chemistry of High-Energy Materials. Berlin, New York: Walter de Gruyter GmbH; 2011

45. Hendrich C, Ciezki H, Schlechtriem S, Gernoth A, Wingborg N, Scharlemann C. Influence of water content in an ADN based liquid mono-propellant on performance characteristics. In: Space Propulsion 2014; Köln. 2014

46. Negri M, Hendrich C, Wilhelm M, Freudenmann D, Ciezki HK, Gediminas L, et al. Thermal ignition of ADN-based propellants, SP2016 paper 3125004. In: Space Propulsion 2016; 2-6 May 2016; Rome, Italy. 2016

47. Larsson A, Wingborg N, Elfsberg M, Appelgren P. Characterization and Electrical Ignition of ADN-Based Liquid Monopropellants; FOI-R-1639-SE; 2005

48. Neff K, King P, Anflo K, Möllerberg R. High performance green propellant for satellite applications. In: 45th AIAA/ASME/SAE/ASEE Joint Propulsion Conference; 2-5 August 2009; Denver, Colorado. 2009

49. Natan B, Perteghella V, Solomon Y. Hypergolic ignition by fuel gellation and suspension of reactive or catalyst particles. Journal of Propulsion and Power. 2011;27:1145

50. Connell TL Jr, Risha GA, Yetter RA, Natan B. Ignition of hydrogen peroxide with gel hydrocarbon fuels. Journal of Propulsion and Power. 2018;34:170

51. Connell TL Jr, Risha GA, Yetter RA, Natan B. Hypergolic ignition of hydrogen peroxide/gel fuel impinging jets. Journal of Propulsion and Power. 2018;34:182

52. Connell TL Jr, Risha GA, Yetter RA, Natan B. Characterization of gelled hydrocarbon fuels and hydrogen peroxide as a hypergolic bi-propellant. International Journal of Energetic Materials and Chemical Propulsion. 2018;17(1):57-78

53. Spicer PT, Gilchrist JF. Ch 3b: Microstructure, rheology, and processing of complex fluids. In: Kresta SM, Etchells AW III, Dickey DS, Atiemo-Obeng VA, editors. Advances in Industrial Mixing: A Companion to the Handbook of Industrial Mixing. 1st ed. John Wiley & Sons, Inc.; 2015. Available from: http://www.nonequilibrium.com/pdf/NAMFMixingRheologyChapter.pdf

54. Beris A, Tsamopoulos J, Armstrong R, Brown R. Creeping motion of a sphere through a Bingham plastic. Journal of Fluid Mechanics. 1985;158:219-244. Available from: https://www.cambridge.org/core/services/aop-cambridge-core/content/view/58DA7E990C0D389D714EC311337CB0E0/S0022112085002622a.pdf/creeping_motion_of_a_sphere_through_a_bingham_plastic.pdf

55. Chhabra RP. Bubbles, Drops, and Particles in Non-Newtonian Fluids. Boca Raton, FL: CRC Press, Taylor & Francis; 2007. Available from: https://www.taylorfrancis.com/books/9781420015386

56. TA Instruments. Rheometers Brochure (RH075B). Available from: http://www.tainstruments.com/pdf/literature/ar2000.pdf [Accessed: July 12, 2017]

57. Ciezki HK, Kirchberger C, Stiefel A, Kröger P, Caldas Pinto P, Ramsel J, et al. Overview on the German gel propulsion technology activities: Status 2017 and outlook. In: EUCASS-2017, the 7th European Conference for Aeronautics and Space Sciences; 3-6 July 2017; Milan, Italy. 2017

2 Chapter

HIGH STRAIN RATE CHARACTERIZATION OF THERMOPLASTIC FIBER-REINFORCED COMPOSITES UNDER COMPRESSIVE LOADING

1. INTRODUCTION: AN OVERVIEW

Aircraft structures experience dynamic loads like collisions with foreign objects, projectile impacts, and shock waves during service, which can have a notable impact on the mechanical characteristics of thermoplastic fiber-reinforced (TFR) composite materials utilized in these advanced frameworks [1, 2, 3]. Therefore, a thorough mechanical analysis at high strain rates is essential for designing composite components effectively. Using static properties for material selection and design due to the absence of dynamic properties can lead to unnecessary structural weight or unexpected failures [1, 4, 5, 6].

Research has been conducted to investigate the correlation between the mechanical properties of fiber-reinforced polymer matrix composites (FRPC) at high strain rates and various material configurations (resin and fiber length, concentration, and orientation) through the utilization of several high strain rate testing methods. Various authors have reviewed and examined numerous research works on this topic. The majority of these studies concentrate on thermoset composites, particularly those with epoxy and polyester matrices reinforced with glass and carbon fibers [3, 5, 7, 8, 9, 10, 11, 12, 13]. Existing literature on thermoplastic composites primarily focuses on polyamide-reinforced composites, ethylene-propylene copolymer matrix reinforced with discontinuous glass fibers, commingled e-glass/polypropylene woven fabric composite, glass fiber-reinforced polypropylene and polybutene-1, and AS4 graphite/polyetheretherketone thermoplastic composite. However, there is limited information available regarding the behavior of PPS matrix composites[4, 11, 14, 15, 16, 17, 18, 19, 20].

The split-Hopkinson pressure bar testing method is commonly utilized for composite materials to achieve high strain rates for tests [21]. This technique involves calculating [3, 5, 10, 18, 22, 23, 24, 25, 26, 27, 28, 29],the specimen stress-time and strain-time response based on the strain waves measured on the bars. High-speed cameras with high resolution can be used to apply optical and contactless strain field measurement techniques like digital image correlation (DIC). This allows for accurate data reduction and provides more information on the distribution of strain across the specimen surface, which is useful for dynamic material characterization[11, 26, 30, 31].. The current study employs these methods to analyze how the strain rate impacts the mechanical properties of PPS matrix carbon fiber-reinforced composite under compressive loads in static and dynamic conditions. Results from dynamic statics are compared to quasi-static test results for the identical specimen geometry and batch. High-speed photography is employed to discover macro-failure modes caused during high strain rate tests, while micro-failure observation is conducted to uncover quasi-static failure characteristics and damage mechanisms of the material across all tested strain rates.

2. METHODOLOGY

2.1. Material

The materials used in this study are matrix polymeric (PPS, polyphenylene sulfide) fiber-reinforced composites, which were selected due to their application on aircraft structures such as leading edges, door structures, and pylon engine covers, among others [32, 33, 34, 35, 36]. These materials were provided by TenCate in the form of rectangular laminates in two versions: 33-ply-thick laminate consisting of carbon fabric, 5HS style (harness-satin weave where one filling yarn floats over four warp yarns and under one, Figure 1a), 3K (3000 individual strands of carbon per fiber bundle) T300J (Toray Carbon Fibers America, Inc. specification, commonly used in aerospace applications), and 280 gsm FAW (fiber area weight in grams per square meter), combined with 42% RC (resin content by weight) Fortron 214 PPS in an orthotropic (0, 90) balanced/mirrored layup (Figure 1a)

42-ply-thick laminate consisting of fiberglass fabric, 8HS style (harness-satin weave where one filling yarn floats over seven warp yarns and under one, Figure 1b), EC6 yarn (e-glass continuous fiber with filament diameter of 6 μm), and 300 gsm FAW (fiber area weight in grams per square meter), combined with 33% RC (resin content by weight) Fortron 214 PPS in an orthotropic (0, 90) balanced/mirrored layup (Figure 1b).

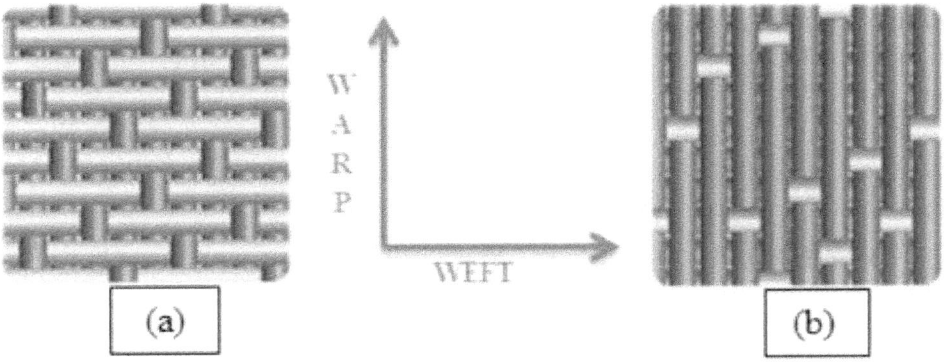

Figure 1. Harness-satin weave (a) 5HS and (b) 8HS [37].

Rectangular specimens of 9 × 10 × 9.8 mm3 (width × length × height) were cut from the originals bars using the Extec Labcut 5000, available on the Lightweight Structures Laboratory of the Institute de Pesquisas Tecnológicas (IPT), which guarantees samples with parallel tolerance of 0.03 mm. The geometry of the specimens was set between the ranges specified to accomplish with the assumptions made about inertia and friction effect [28]. Figure 2 shows the specimen configuration; red arrows indicate the load direction.

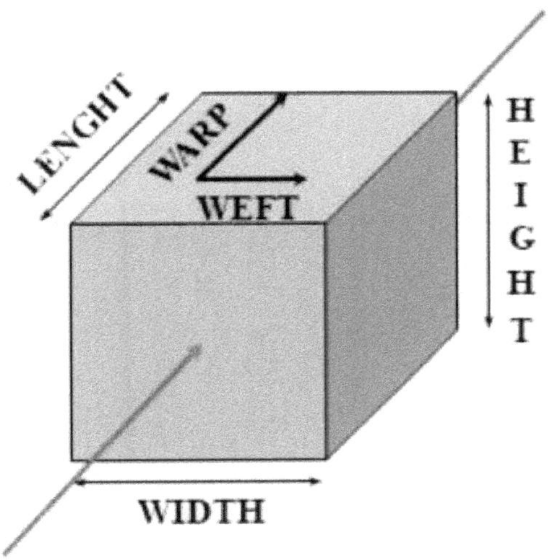

Figure 2. Isometric view of the specimen.

2.2. Quasi-Static Test

Quasi-static reference tests were performed at the Lightweight Structures Laboratory (IPT), using an INSTRON servo-mechanic universal testing machine at three constant displacement rates (0.6, 6, and 60 mm/min), which corresponds to three quasi-static axial strain rates 0.001, 0.01, and 0.1 s−1 by considering the nominal specimen length of 10 mm. Load measurements were recorded using universal testing machine hardware and software, while an Imetrum video gauge system synchronized with the testing machine was used to measure strain in the specimen. Using the software interface, target points were placed at the center of the specimen, and the strain measurements were recorded during the test; additional targets were set to monitor strain behavior.

Special testing fixtures, with load distribution system, were used in the testing machine movable head member in order to ensure alignment during specimen loading process. Figure 3 represents a schematic setup for quasi-static tests.

Longitudinal compressive strength is calculated, for this regimen, according to expression σ_S = P/AS, where P is the applied load recorded by universal testing machine hardware and software and AS is the cross-sectional area of the specimen [38].

Strength and strain data are post-processed to build up the stress-strain curves for quasi-static regime and to calculate peak stress, strain at peak stress, and Young's modulus in order to compare mechanical properties with those obtained in the dynamic regime. Experimental Young's modulus is determined as the slope of the linear regression (LR) applied to the stress-strain curve in the range of strain data between 0.7 and 1%.

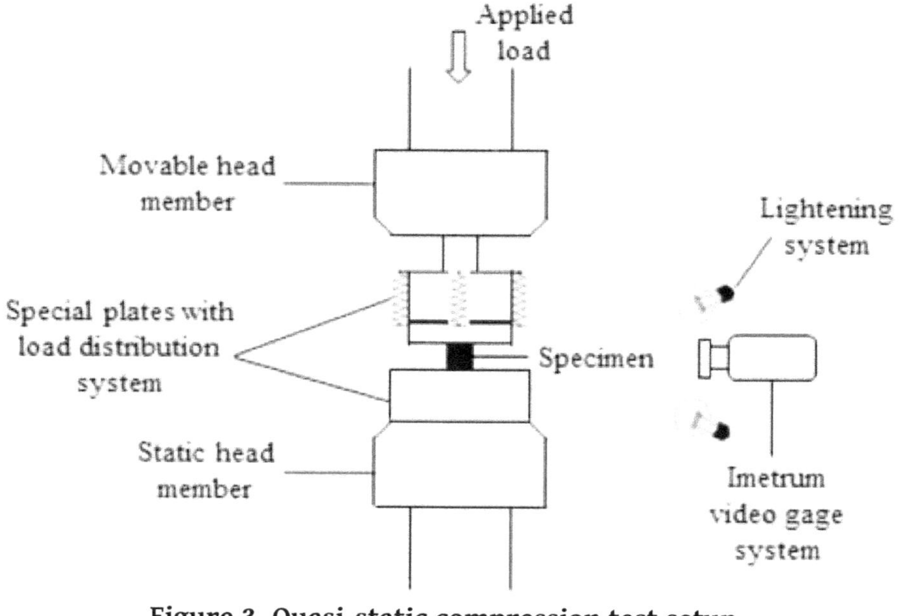

Figure 3. Quasi-static compression test setup.

2.3. Dynamic Test

High strain rate tests were performed using a split-Hopkinson pressure bar apparatus available in the Aerospace Structures Laboratory at Instituto Tecnológico de Aeronáutica (ITA) composed by three cylindrical bars: striker, incident, and transmitted bar. All three bars are made of high strength steel AMS 5629 with Young's modulus of 198 GPa, density of 7700 kg/m3, yield stress of 1.4 GPa, and diameter of 19.05 mm. Striker length is 350 mm, and incident/transmission bars have 1000 mm, with length/diameter ratio of 50.0, that ensures the validity of unidimensional wave propagation assumption. The strain measurement system has four strain gauges HBM, model LY11-3/350 with 3 mm grid enabling measurements up to 100 kHz, which are disposed diametrically opposed in order to compensate bending and are located at 50 cm of the contact edge between specimen/incident bar and specimen/transmitted bar, respectively, and a data acquisition and conditioning system HBM Genesis 7 t, with a 16 bit resolution analog/digital card, four strain gauge channels, and sampling rate of 1 MHz. The present configuration has a momentum trap after the transmission bar to preserve the strain gauges [29, 39]. Additionally, a Photron high-speed imaging system (HSIS) composed of a high-speed camera model FASTCAM SA-Z and FASTCAM analysis software was set up with the SHPB apparatus to capture videos and images during the experiments. Figure 4 shows the schematic SHPB testing setup used for the dynamic tests.

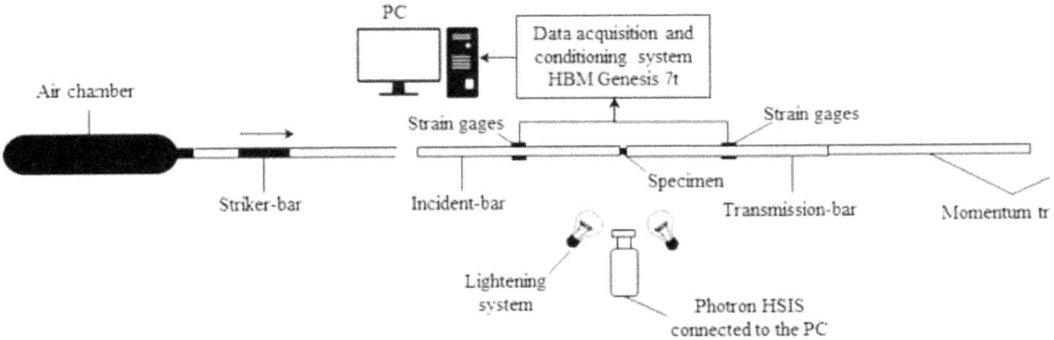

Figure 4. Schematic SHPB apparatus and HSIS set up.

Data acquisition and conditioning system signals were post-processed using an in-house Python program, which computes the stress, strain, and strain rate on the specimen using the classical SHPB analysis based on the one-dimensional wave propagation theory, which implies elastic deformation in the bars during the tests, unidirectional elastic pulses propagate along the bars, uniform deformation process in the specimen, and no dispersion of waves throughout the bars and the specimen [11, 24, 26, 28, 29].

Tests were done at two different pressure values in the air chamber, 1.2 and 1.6 bar, which correspond to two different strain rates for each material (558.5 and 891.1 s-1 for glass fiber, and 400.5 and 832.2 s-1 for carbon fiber).

Stress-strain database were used to build up the stress–strain curves for specimens tested in dynamic regime and to calculate mechanical properties (peak stress, strain at peak stress, and Young's modulus). Experimental Young's modulus is determined as the slope of the linear regression (LR) applied to the stress-strain curve in the range of strain data between 0.7 and 1%.

2.4. Digital Image Correlation

Digital image correlation (DIC) technique was used to measure strain during dynamic tests using the HSIS images obtained. 2D deformation vector fields and strain maps were built for specimens at each high strain rates tested, obtaining strain measures in an overall area and along an inspection line (L0) at each deformation state (Figure 5). Strain data from the center portion of the inspection line (L0) was synchronized with stress data from SHPB apparatus, and stress-strain curves were built using strain measured by DIC in order to compare the DIC-based results with the results obtained from SHPB strain gauge measurements and quasi-static regime results.

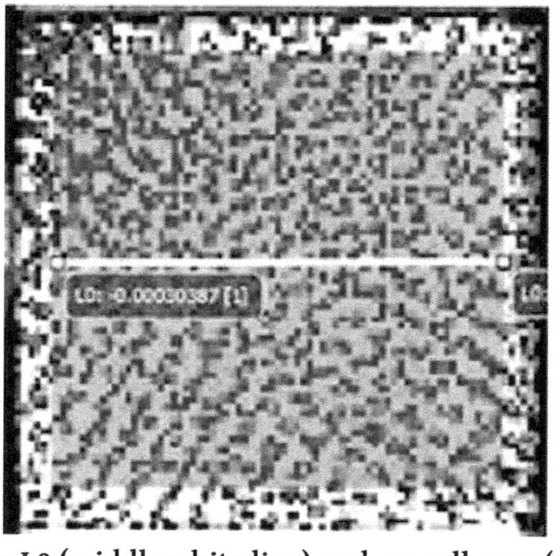

Figure 5. Inspect line L0 (middle white line) and overall area (purple square) used for DIC measurements.

3. EXPERIMENTAL RESULTS AND DISCUSSION

3.1. PPS Carbon Fiber-Reinforced Composite

3.1.1. Quasi-Static Tests

Quasi-static tests are performed to failure, under the three deformation rates previously specified (0.001, 0.01, and 0.1 s-1). Figure 6 shows stress-strain curves obtained for each strain rate applied, while Table 1 summarizes the post-processed results. The results obtained show that material's peak stress remains constant under quasi-static regime with an average value of 532.603 MPa; this is based on the low value of the standard deviation (std. dev.) and the coefficient of variation (CV). The same trend is observed for the Young's modulus obtaining an average value of 43.859 GPa. On the other hand, the strain at peak stress shows an average of 1.284% with a variation coefficient of 22.45%, which indicates that it varies considerably; however, it does not show a sensitivity on the strain rate or a correlation with a variation in the failure modes observed by fractographic observation, which indicates that dispersion in the measurements can be attributed to intrinsic errors in the measurement method. Having this in mind, it can be concluded that the mechanical behavior of the material is not strain rate sensitive within the strain rate range tested in the quasi-static regime.

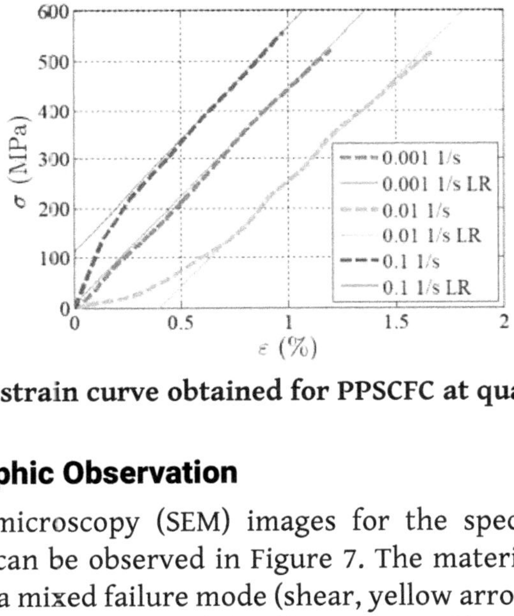

Figure 6. Stress-strain curve obtained for PPSCFC at quasi-static regime.

3.1.1.1. Fractographic Observation

Scanning electron microscopy (SEM) images for the specimens tested in the quasi-static regime can be observed in Figure 7. The material tested at 0.001 s-1 (Figure 7a) presents a mixed failure mode (shear, yellow arrows; delamination, red arrows) due to the configuration of the fiber. Woven fiber laminates commonly present delamination through the warp fibers and shear through the weft fibers. Additionally, local kinkbands (Figure 7a, discontinued white lines) are evidenced in the warp fiber bundles, which is common in compression load failure due to the microbuckling that is developed in the fibers aligned in the direction of the load [40]. The specimens tested under 0.01 s-1 (Figure 7b) and 0.1 s-1 (Figure 7c) present similar behavior, which is observed a mixed failure mode (delamination and shear) and the development of local kinkbands. The material tested at 0.01 s-1 develops intralaminar failure (discontinued yellow lines) related to the failure of the weft fibers and microbuckling (discontinued red line).

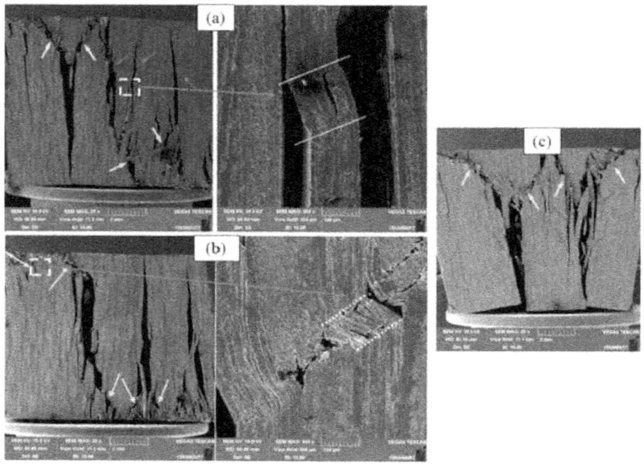

Figure 7. Quasi-static failure modes observation by SEM for PPSCFC under (a) 0.001 s-1 (at 25× with amplified zone at 500×), (b) 0.01 s-1 (at 25× with amplified zone at 500×) and (c) 0.1 s-1.

3.1.2. Dynamic Tests

The dynamic tests are performed up to failure under two average strain rates of 400.5 and 832.3 s-1; all these tests achieved equilibrium forces on specimen's surfaces, which is evidenced in Figure 8 where forces applied on each contact surface of the specimen are showed for a test under each strain rate.

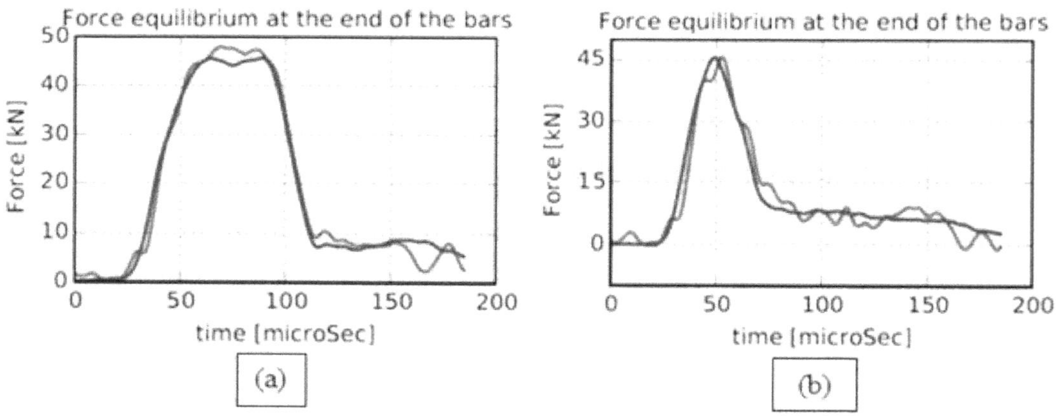

Figure 8. Force equilibrium on the edge of the incident bar (red line) and transmitted bar (blue line) for PPSCFC under (a) 400.5 s-1 and (b) 832.3 s-1.

Stress-strain curves and data obtained for each strain rate applied at dynamic regime (Figure 9 and Table 2) evidenced that the mechanical behavior of the material is not significantly affected by the strain rate effects presenting an average strength of 530.237 MPa, failure strain of 2.141%, and elasticity modulus of 34.273 GPa when strain is measured by SHPB strain gauges.

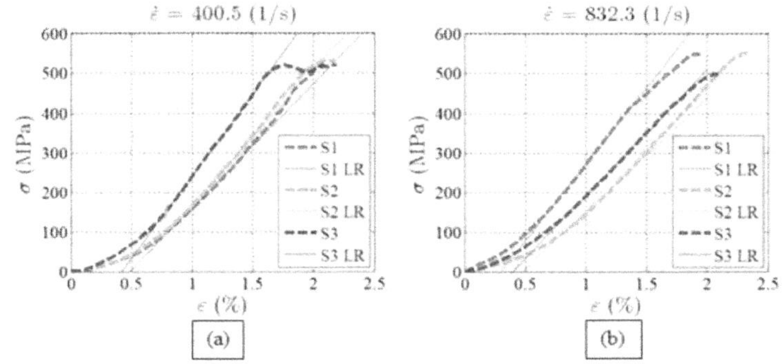

Figure 9. Stress-strain curves obtained under dynamic regime for PPSCFC. (a) 400.5 s-1 and (b) 832.3 s-1.

The same behavior remains when strain is measured by DIC; however the strain measured on the center portion of the specimen by DIC is lower than SHPB strain of about 38.856 and 36.605% for 400.5 and 832.3 s-1, respectively, which leads to an increase in the Young's modulus of about 90.65% for 400.5 s-1 and 71.606% for 832.3 s-1. This difference can be seen in Figure 10, where stress-strain curves with strain measured by SHPB system and by DIC are shown at each tested strain rate.

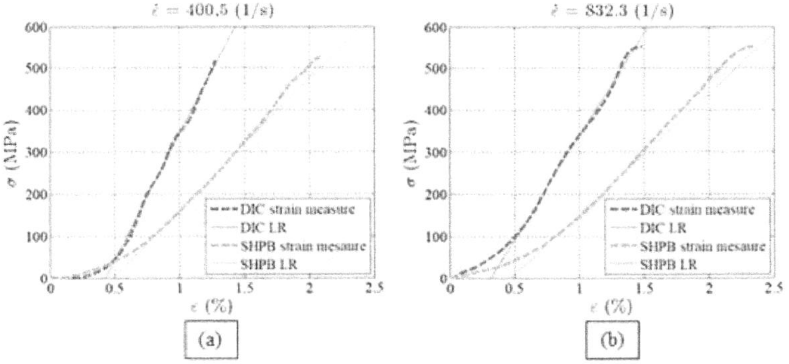

Figure 10.Stress-strain curve obtained for PPSCFC with strain measured by DIC and SHPB strain gages under (a) 400.5 s-1 and (b) 832.3 s-1

Table 1. Experimental results for PPSCFC at quasi-static regime.

Strain rate (s^{-1})	Peak Stress (MPa)	Strain at peak stress (%)	Young Modulus (GPa)
0.001	521.926	1.203	43.664
0.01	517.869	1.67	42.271
0.1	558.013	0.978	45.643
Average	532.603	1.284	43.859
Std. Dev.	18.04	0.288	1.383
CV (%)	3.388	22.45	3.155

The strain map for the overall area at the moment of the failure obtained by DIC strain analysis indicates that strain value measured by strain gauges was reached in localized points of the specimen, generally on the edges (2.25% in the DIC spectrum and 2.088% according to SHPB measure for 400.5 s-1; and 2437% on the DIC spectrum and 2.346% according to SHPB measure for 832.3 s-1). This indicates that dynamic test behaves according to the theory and generates deformation highly localized or not homogeneous along the specimen due to the test high speed [24, 25, 26, 28, 41, 42]; also, the value measured by strain gauges is a real value of strain within the specimen; however, this value is the highest reached in all the specimen; in consequence, it is wise to measure strain by the DIC technique to obtain an accurate value on the center of the specimen where it can be assured that the behavior of the material is not influenced by the effect of the edges, where higher tendency to failure can be present as a consequence of the specimen machining [43].

Table 2. Comparison of mechanical properties under high strain rates for PPSCFC.

Strain rate (s^{-1})	Peak Stress (MPa)	Strain at peak stress (%)	Young Modulus (GPa)
400.5	525.747	2.148	35.493
832.3	534.726	2.134	33.053
Average	530.237	2.141	34.273
Std. Dev.	4.489	0.0073	1.22
VC (%)	0.847	0.341	3.559

3.1.2.1. Fractographic Observation

The fractographic observation for dynamic regime is performed by monitoring the failure through the high-speed image system and SEM post-failure observation. Figure 11 shows the sequence of the material's failure submitted to 400.5 s-1 where the red arrow indicates the direction of the compressive wave. It can be observed the specimen at the beginning of the failure without any failure indication in Figure 11a; in the next image (Figure 11b) is shown the beginning of the failure on transmitted bar/specimen edge (blue circle), which is attributed to material's edges weakening by machining effect. The beginning of the failure is given in the form of delamination, and then it is propagated diagonally or in shear mode (Figure 11c). Besides, other two crack fronts are initiated on incident bar/specimen edge (blue circle Figure 11c), which join in a "v" shape, and then it is propagated in delamination mode. The specimen finishes its failure process with a partial separation of the surfaces (Figure 11d).

The material's failure process observed by HSIS for strain rate of 832.3 s-1 is shown in Figure 12. The beginning of the failure is observed in two regions of the transmitted bar/specimen edge (blue circle) taking as reference the direction of wave propagation (red arrow) (Figure 12b). The upper region is submitted to bending, which is why the material looks bended upward, while the lower region develops two delamination fronts. The delaminations are propagated, while the bending on the upper region is intensified generating delamination and separation of the plies (Figure 12c) indicating failure of the resin, until the failure of the fibers under shear is initiated (Figure 12d yellow line). The propagation of the different crack fronts is prolonged until the specimen is entirely divided in several pieces (Figure 12e).

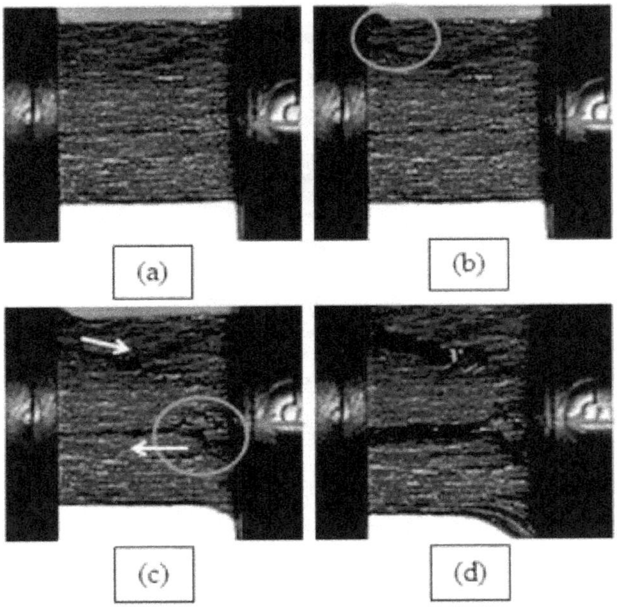

Figure 11. Images sequence taken by HSIS for PPSCFC tested at 400.5 s^{-1}.

SEM fractographic observation was performed on the fracture surface of the different recovered parts of the material tested under 832.2 s–1. Figure 13a shows a surface that appears to be "melted," while Figure 13b–d presents other surface that is "unmelted." This behavior indicates that heat generation during the test may have affected the crystallinity degree of the thermoplastic matrix. The "melted" surface (Figure 13a) is characterized for being smooth and without any distinctive features; observation at high zooms show the presence of fibrils in the interior of a crack, which look like resin threads that try to keep the crack faces together and oppose the propagation. On the "unmelted" surface (Figure 13b), two zones can be observed, Z1 which seems to be a zone directly over the fabric (it is not possible to identify if it is warp or weft) and Z2 which seems to be an interstitial site or a high resin content site. The zone Z1 (Figure 13c) shows cusps and scallops (red arrows), while the zone Z2 (Figure 13d) presents riverlines (yellow arrows) and feather marks (red circles). This fractographic aspects are typical of a composite submitted to compression [40, 44, 45, 46].

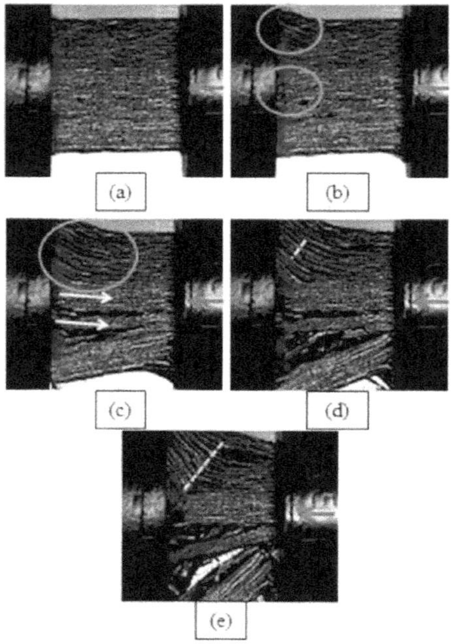

Figure 12. Images sequence taken by HSIS for PPSCFC tested at 832.2 s^{-1}.

3.1.3. Mechanical property comparison for PPSCFC

Tables 3 and 4 present that the mechanical properties are not significantly affected by the strain rate effect while strain rate increases. Peak stress variation is less than 3.5% when results of specimens tested at quasi-static and dynamic regimes are compared, which indicate that strength is not strain rate dependent for this material. Strain at peak stress measured by the SHPB system presents higher variation than the strain measured by DIC (29.23–17.977%, respectively) compared with quasi-static results; however, the failure mechanism has not presented any difference which is why both can be considered negligible and the difference is attributed to the measurement method. The variation observed on the modulus is given by the obtained strain data; higher values for strain at peak stress obtained by SHPB gauges give modulus' lower values for dynamic tests and lower variation between quasi-static and dynamic regime compared with results with strain data measured by DIC; however, modulus is taking as constant, too, that is, it is not strain rate dependent.

Strain rate insensitivity is also observed in Figure 14, where peak stress-strain rate relation (a), strain at peak stress-strain rate relation (b), and Young's modulus-strain rate relation (c) are shown. Quasi-static (QS) and dynamic (D) average data is plotted for each property, differentiating SHPB strain data and DIC strain data.

It can be concluded that the material is not strain rate dependent and the measurement of dynamic strain by DIC allows a more accurate comparison with respect to the measurement of the quasi-static strain by the video strain gauge system. PPSCFC mechanical behavior discussed does not coincide with the behavior observed on similar materials (thermoplastic matrixes reinforced with carbon fiber) [11, 16, 19, 23, 27, 38]. This can be explained by the effect of carbon fiber on the resin (PPS) crystallization reported in open literature, where it has been found that transcrystallinity on the fiber-resin interphase affects the mechanical properties of the composite [32, 38, 47, 48, 49].

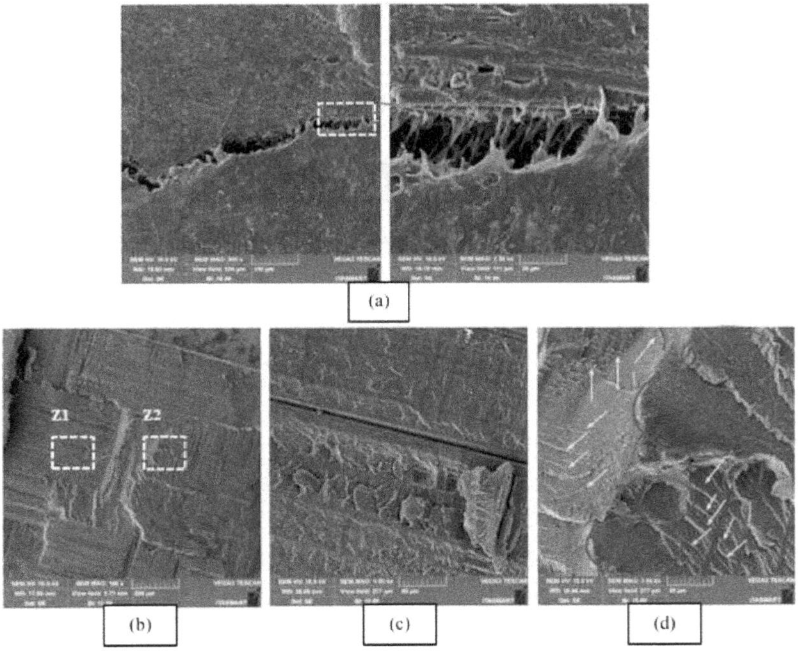

Figure 13. SEM images for PPSCFC tested at 832.3 s^{-1}. (a) "Melted" surface at 500× with 2500× zoom of the marked zone, (b) "unmelted" surface at 100×, (c) Z1 marked in (b) at 1000×, and (d) Z2 marked in (b) at 1000×.

Table 3.Average mechanical properties for PPSCFC with strain measured by SHPB system.

Strain rate (s^{-1})	Peak Stress (MPa)	Strain at peak stress (%)	Young Modulus (GPa)
0.001	521.926	1.203	43.664
0.01	517.869	1.67	42.271
0.1	558.013	0.978	45.643
400.5 SHPB	525.747	2.148	35.493
832.3 SHPB	534.726	2.134	33.053
Average	531.6564	1.6267	40.025
Std. Dev.	14.309	0.475	4.878
CV (%)	2.691	29.23	12.189

Failure mode is not strain rate dependent, either. The material presents mixed failure aspects (delamination and shear), which is characteristic for a laminate fabric submitted to compression according to Greenhalgh [40]. The material tested under the highest strain rate presents separation on multiple parts indicating severe damage due to the high-speed load application and insufficient dissipation of the heat generated in the deformation process. It is also worth to mention that it is observed that the delamination becomes more predominant at highest strain rate, which coincides with what Greenhalgh [40] reports for laminates submitted to high-speed impact loading.

Table 4. Average mechanical properties for PPSCFC with strain measured by DIC.

Strain rate (s^{-1})	Peak Stress (MPa)	Strain at peak stress (%)	Young Modulus (GPa)
0.001	521.926	1.203	43.664
0.01	517.869	1.67	42.271
0.1	558.013	0.978	45.643
400.5 DIC	517.342	1.203	59.765
832.3 DIC	552.518	1.487	49.263
Average	533.534	1.323	48.121
Std. Dev.	17.899	0.238	6.279
CV (%)	3.355	17.977	13.0479

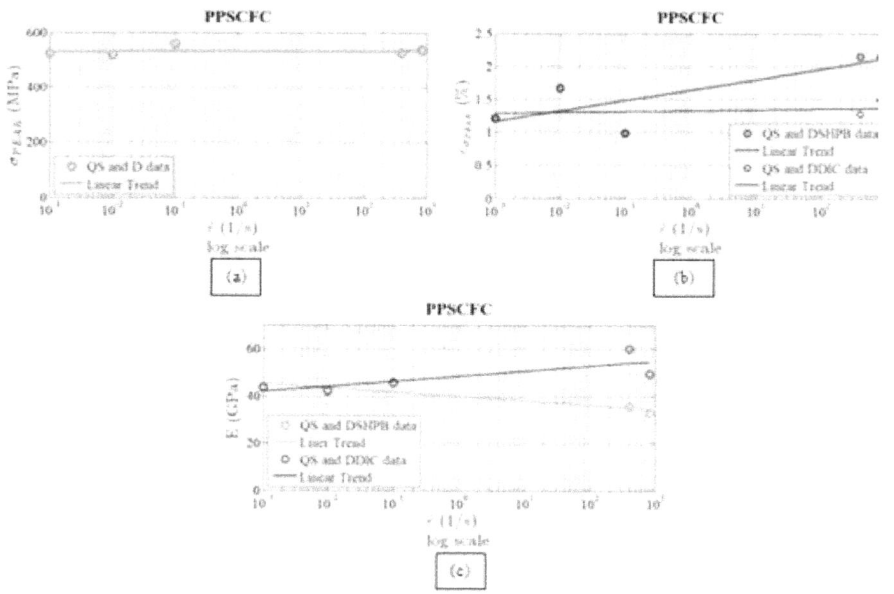

Figure 14. PPSCFC properties as function of the strain rate: (a) strength-strain rate plot, (b) ultimate strain-strain rate plot and (c) Young's modulus-strain rate plot.

3.2. PPS Glass Fiber-Reinforced Composite

3.2.1. Quasi-Static Tests

Quasi-static tests are performed to failure, under the three deformation rates specified (0.001, 0.01, and 0.1 s-1). Figure 15 shows stress-strain curves obtained for each strain rate applied, while Table 5 summarizes the post-processed results. Sample S1 for 0.001 s-1 tests was discarded due to reload during the test. The average value obtained for the peak stress for the quasi-static regime is 358.295 MPa with a coefficient of variation (CV) of 5.235%, which means that it is constant. Strain at peak stress and modulus presents the same behavior with values as of 1.676%-(CV) 8.875% and 21.999 GPa-(CV) 2.623%, respectively. Due to this it is said that mechanical behavior remains constant in quasi-static regime.

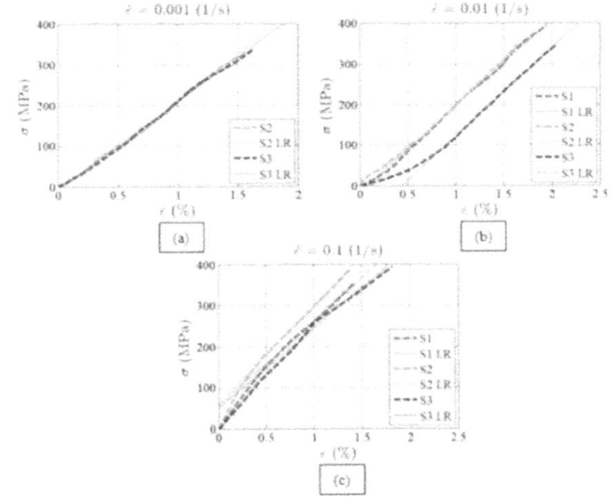

Figure 15. Stress-strain curve obtained for PPSGFC under compression at (a) 0.001 s-1, (b) 0.01 s-1 and (c) 0.1 s-1.

Table 5. Experimental results for PPSGFC at quasi-static regime.

Strain rate (s^{-1})	Peak Stress (MPa)	Strain at peak stress (%)	Young Modulus (GPa)
0.001	333.421	1.606	22.786
0.01	362.757	1.883	21.79
0.1	378.708	1.539	21.419
Average	358.295	1.676	21.999
Std. Dev.	18.756	0.149	0.577
CV (%)	5.235	8.875	2.623

3.2.1.1. Fractographic Observation

Fractographic observation by SEM for the PPSGFC specimens tested under quasi-static regime evidences a mixed failure mode (delamination and shear) for the laminate submitted to compression according to what is reported by Greenhalgh for fabrics [40]. Figure 16 shows the failure modes for the three strain rates identifying delamination with red arrows and the shear with yellow arrows. The material tested at 0.001 s-1 developed an early stage kinkband (Figure 16b), which indicates basic failure mode for composites under compression [38, 40, 44]. Figure 16e indicates the formation of a fiber bridging characteristic of mode I delamination (opening) [44].

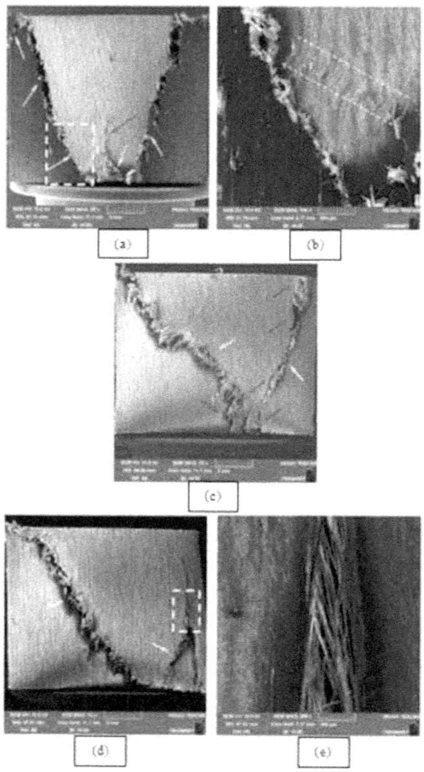

Figure 16. SEM images for PPSGFC tested under quasi-static strain rates. (a) 0.001 s-1 at 25×, (b) marked zone in (a) at 100×, (c) 0.01 s-1 at 25×, (d) 0.1 s-1 at 25× and (e) marked zone in (d) at 250×.

3.2.2. Dynamic Tests

The dynamic tests are performed up to failure under two average strain rates of 558.5 and 891.1 s-1; all these tests achieved equilibrium forces on specimen's surfaces which is evidenced in Figure 17 where forces applied on each contact surface of the specimen are showed for a test under each strain rate.

According to the results (Figure 18 and Table 6), material's mechanical behavior is constant for the dynamic regime presenting 491.554 MPa as peak stress, 2.647% as failure strain, and 22.498 GPa as Young's modulus when strain is measured by SHPB strain gauges.

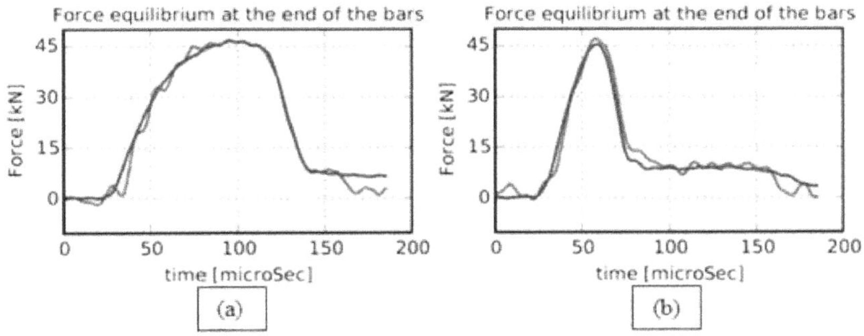

Figure 17. Force equilibrium on the edge of the incident bar (red line) and transmitted bar (blue line) for PPSGFC under (a) 558.5 s-1 and (b) 891.1 s-1.

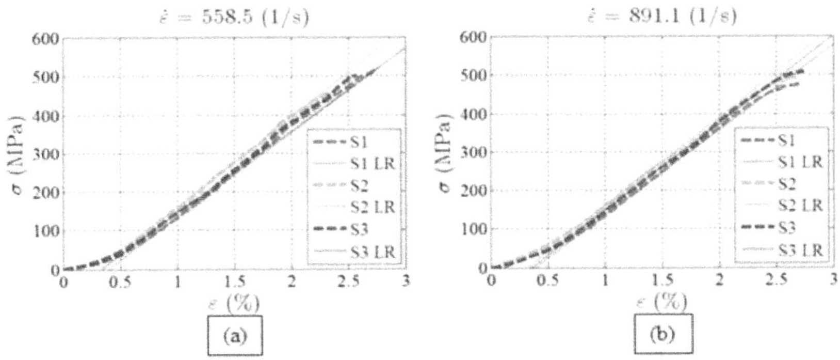

Figure 18.Stress-strain curves obtained under dynamic regime for PPSGFC: (a) 558.5 s-1 and (b) 891.1 s-1.

Table 6. Comparison of mechanical properties under high strain rates for PPSGFC.

Strain rate (s^{-1})	Peak Stress (MPa)	Strain at peak stress (%)	Young Modulus
558.5	491.886	2.575	22.293
891.1	491.223	2.719	22.703
Average	491.554	2.647	22.498
Std. Dev.	0.332	0.072	0.205
CV (%)	0.067	2.7322	0.912

For this material it is also performed the strain measurement processing the high-speed videos under DIC, as it was done for the PPSCFC. Figure 19 shows the variation of the stress-strain curves built up with the strain data measured by SHPB and DIC. Results obtained for this material are similar to PPSCFC; the strain measured by DIC is lower in 25.495 and 19.627% for 558.5 and 891.1 s-1, respectively, with respect to the deformation measured with the SHPB strain gauges. As it was expected, the peak stress remains equal what leads to modulus increase (38.054%

for 558.5 s-1 and 6.553% for 891.1 s-1) with respect to the data obtained by the SHPB.

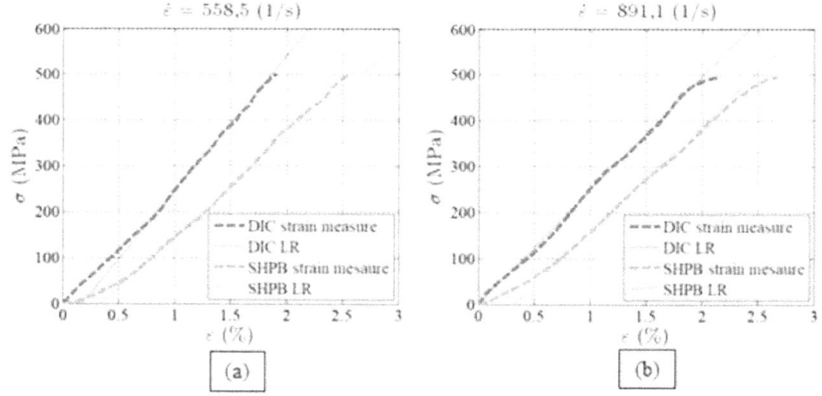

Figure 19. Stress-strain curve obtained for PPSGFC with strain measured by DIC and SHPB strain gages under (a) 558.5 s-1 and (b) 891.1 s-1.

The behavior observed in the strain map obtained by DIC is the same to the one analyzed for PPSCFC; the highest strain is very close to the value measured by the SHPB system (2.437% DIC–2.573% SHPB at 558.5 s-1; 2.843% DIC–2.68% SHPB at 891.1 s-1), which reaffirms what was previously said; the SHPB system obtains the highest strain value in the entire specimen. Resulting strain in the specimen is not homogeneous along it and is more critical on the edges, and the measurement of the strain DIC allows obtaining data without the influence of the machining.

3.2.2.1. Fractographic Observation

The same that for PPSCFC, failure process monitoring is performed for PPSGFC through the high-speed image system; additional SEM post-failure observation was realized.

Images for material tested at 558.5 s-1 (Figure 20) indicate that the failure starts on the specimen/transmitted bar contact edge and it propagates as delamination (Figure 20b and c, white arrow). A second crack front is developed on the inferior part of the specimen/incident bar contact edge in shear (Figure 20c, blue circle), which propagates and meets other crack front forming a "v" shape and continuing the propagation as delamination (Figure 20d), resulting in a mixed failure mode (delamination and shear).

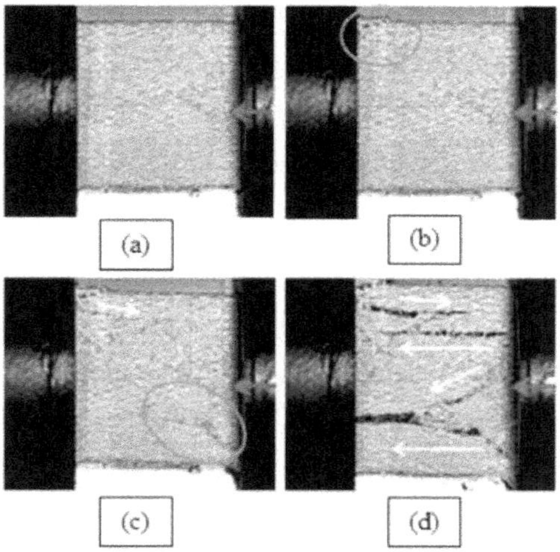

Figure 20. Images sequence taken by HSIS for PPSGFC tested at 558.5 s^{-1}.

Failure behavior for the material tested at 891.1 s–1 is similar to that observed in specimens tested in other strain rates. The beginning of the failure starts on the inferior region of the specimen/transmitted bar contact edge, and it propagates in shear (Figure 21b, blue circle); on its way it is bifurcated; and on one side, it continues in shear toward the superior opposite edge; and on the other side, it propagates as delamination forming a "v" shape in the material (Figure 21c, white arrows). Additionally, another crack front is developed on the superior part of the specimen/transmitted bar contact edge in form of delamination, which is propagated separating the part and allowing relative movement with respect to the other parts (Figure 21d). For this strain rate, the material is divided in several parts which were submitted to SEM observation on the fracture surface.

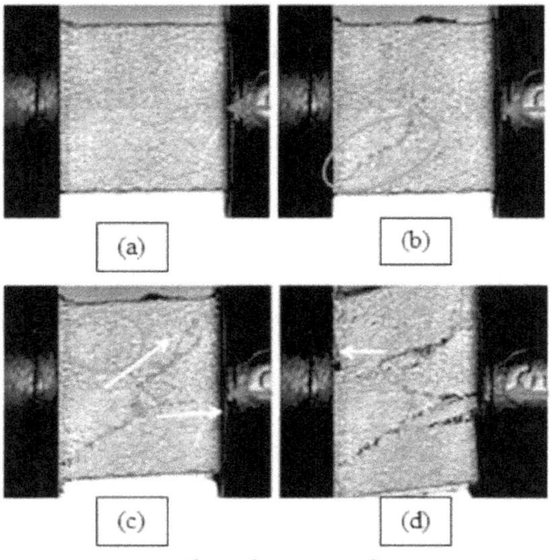

Figure 21. Images sequence taken by HSIS for PPSGFC tested at 891.1 s^{-1}.

The analysis of the fracture surface for the material tested at 891.1 s-1 indicates the development of two types of surface as in the PPSCFC on the highest strain rate. A surface appears "melted" (Figure 22a), while the other appears "unmelted" (Figure 22b). The "melted" surface exhibits signs of abrasive wearing possibly due to the movement between sheets observed on the HSIS images [50, 51], while the "unmelted" surface presents cusps (red circle) which appears to be the weft as in mode I (opening) [44].

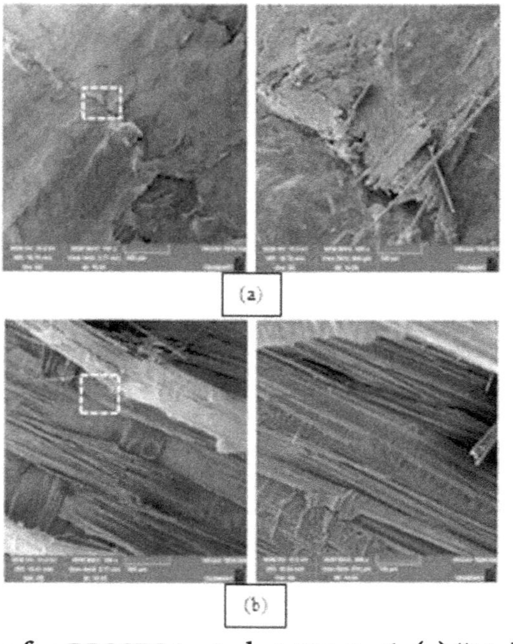

Figure 22. SEM images for PPSGFC tested at 891.1 s^{-1}. (a) "Melted" surface at 100× with 500× zoom of the marked zone, (b) "un-melted" surface at 100× with 500× zoom of the marked zone.

3.2.3. Mechanical Property Comparison for PPSGFC

The data obtained by the SHPB system (Table 7) evidence an increase of 27% on the peak stress when the average values for the quasi-static and dynamic regimes are compared (358.295 and 491.554 MPa). Similarly, the deformation on the ultimate stress is increased by 36% when the average values for the quasi-static and dynamic regimes are compared (1.676 and 2.647%). It was not evidenced a significant change in the Young's modulus for the material tested in the dynamic regime with respect to that of the semi-static regime (2% increase).

According to the data obtained when strain is measured by DIC (Table 8), the tendency of strength increase is kept at 38% when the averages of the quasi-static and dynamic regimes are compared (358.295 and 496.758 MPa), which is expected given that the determination of the peak stress is made with the same data.

However, the increase in the strain at peak stress is just 21% based on the average values for each regime (1.676% quasi-static and 2.034% dynamic).

Table 7. Average mechanical properties for PPSGFC with strain measured by SHPB system.

Strain rate (s^{-1})	Peak Stress (MPa)	Strain at peak stress (%)	Young Modulus (GPa)
0.001	333.421	1.606	22.786
0.01	362.757	1.883	21.79
0.1	378.708	1.539	21.419
558.5 SHPB	491.886	2.575	22.293
891.1 SHPB	491.223	2.719	22.703
Average	411.599	2.065	22.198
Std. Dev.	66.881	0.492	0.526
CV (%)	16.249	23.822	2.369

The aforementioned affects the behavior of the Young's modulus, which presents an increase of 23% according to the average values for quasi-static and dynamic regimes (21.999 and 27.168 GPa). The behavior previously described is graphically evidenced in Figure 23 where the properties' tendency with respect to the strain rate is shown.

Results indicate that the material is strain rate dependent presenting an increase in the properties when strain rate applied increases, which is evidenced more accurately with DIC strain measurement and comparing it with the quasi-static results. The mechanical behavior observed on this material coincides with data reported for other similar materials in open literature (thermoplastic matrixes reinforced with glass fibers) [4, 16, 17, 18, 38, 52, 53].

Table 8. Average mechanical properties for PPSGFC with strain measured by DIC.

Strain rate (s^{-1})	Peak Stress (MPa)	Strain at peak stress (%)	Young Modulus (GPa)
0.001	333.421	1.606	22.786
0.01	362.757	1.883	21.79
0.1	378.708	1.539	21.419
558.5 DIC	494	2.154	24.781
891.1 DIC	499.517	1.914	29.556
Average	413.681	1.819	24.067
Std. Dev.	69.393	0.223	2.982
CV (%)	16.774	12.275	12.392

The fractographic observation performed in both regimes indicates that the failure mechanism is not strain rate dependent since there is no significant variation in the failure aspects observed in specimens tested at high strain rates in comparison to those tested quasi-statically. The failure mechanism for the material is mixed (delamination and shear) in all the strain rates with a notorious dominance of the shear mode, which in the high deformation rates becomes balanced with the delamination. This behavior is expected due to the tendency of the glass fiber to fail at 45° when submitted to compression, and the laminate dilation turns the material more susceptible to delamination at high strain rates, according to Greenhalgh [40].

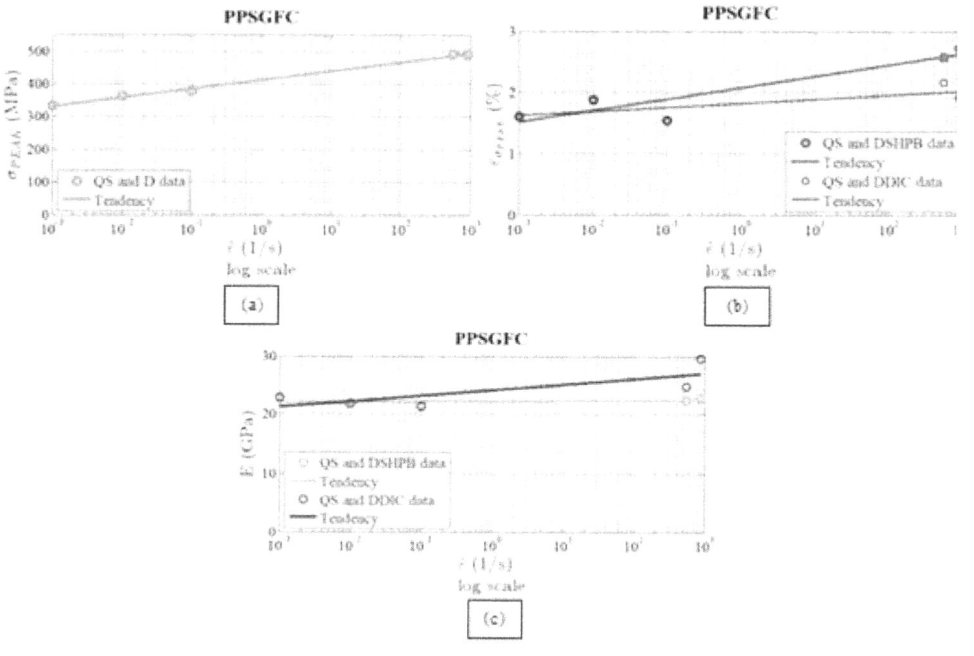

Figure 23.PPSCFC properties as function of the strain rate: (a) strength-strain rate plot, (b) ultimate strain-strain rate plot and (c) Young's modulus-strain rate plot.

4. CONCISE SUMMARY

The materials exhibit linear elastic behavior at all strain rates measured. Mechanical properties are consistent regardless of the strain rate applied in each regime. PPSCFC exhibited a compressive strength of 532.603 MPa, failure strain of 1.284%, and Young's modulus of 43.859 GPa under quasi-static conditions. At high strain rates, the compressive strength was 534.93 MPa, failure strain was 1.345%, and Young's modulus was 53.014 GPa. The PPSGFC test results show the following properties: - Quasi-static regime: Compressive strength of 358.295 MPa, failure strain of 1.676%, and Young's modulus of 21.999 GPa. - Dynamic regime: Compressive strength of 496.758 MPa, failure strain of 2.034%, and Young's modulus of 27.168 GPa. When comparing the mechanical properties calculated under quasi-static and dynamic conditions, it is observed that the PPSCFC shows strain rate insensitive behavior, while the PPSGFC displays strain rate dependency, resulting in improved mechanical properties with higher strain rates. The compressive strength, failure strain, and Young's modulus rise by 38%, 21%, and 23%, respectively. Based on the strain spectrum from DIC, the SHPB system recorded the highest strain value in the specimen. However, due to the dynamic strain behavior, it is advisable to conduct localized strain measurements to compare mechanical properties in both testing conditions. The behavior of the PPSCFC described here does not align with data and reports from other sources. The interaction between the semicrystalline resin (PPS) and carbon fibers during cooling causes transcrystallinity at the fiber-resin interface, impacting the material's mechanical properties. Additional research is needed to determine if this is the cause of the material's altered behavior. The behavior of the PPSGFC aligns with expectations based on data and studies in the literature. The strain rate dependency of the mechanical characteristics is due to the resin's viscoelastic nature. The failure mode observed for the materials is generally heterogeneous. Delamination and shear mode have been established, and it has been observed that the failure aspects are not significantly influenced by strain rates. This leads to the conclusion that the failure mechanism is not dependent on strain rate. The material configuration and the fiber-resin interphase have a greater influence on the failure mechanisms than the applied strain rate. This explains the typical fractographic characteristics observed in composite materials under uniaxial compression. More significant efforts are required to comprehend the formation and dissipation of heat during the material strain process at high strain rates to gain a better understanding of their impact on the fractographic characteristics of the material, particularly the "melted" surfaces.

5. AT A GLANCE

An experimental investigation was conducted on the mechanical properties of polyphenylene sulfide (PPS)-based composite laminates reinforced with carbon and glass fibers. The study examined how the materials responded to compression loads at various strain rates. Quasi-static tests were conducted with an electromechanical universal testing machine at three distinct strain rates, while dynamic tests were performed using a split-Hopkinson pressure bar (SHPB) apparatus at two pressure configurations within the gas chamber. An advanced imaging system was utilized to observe the failure process in real-time during a dynamic test. The images captured were then analyzed using digital image correlation (DIC) to quantify strain. This was done to compare the strain measurements obtained by DIC with those from SPHPB strain gauges and quasi-static tests. Fractography analysis was conducted to determine the primary failure causes that occurred at various strain rates.

REFERENCES

1. Hosur MV, Alexander J, Vaidya UK, Jeelani S, Mayer A. High strain rate compression characterization of affordable woven carbon/epoxy composites under off-axis loading. Polymers and Polymer Composites. 2003;11(7):527-539
2. Federal Aviation Administration. Pilot's Handbook of Aeronautical Knowledge. Oklahoma: United State Department of Transportation; 2016
3. Hosur MV, Alexander J, Jeelani S, Vaidya UK, Mayer A. High strain compression responses of affordable woven carbon/epoxy composites. Journal of Reinforced Plastics and Composites. 2003;22(3):271-296
4. Fitoussi J, Bocquet M, Meraghni F. Effect of the matrix behavior on the damage of ethylene-propylene glass fiber reinforced composite subjected to high strain rate tension. Composites Part B: Engineering. 2012;45:1181-1191
5. Hamouda AMS, Hashmi MSJ. Testing of composite materials at high rates of strain: Advances and challenges. Journal of Materials Processing Technology. 1998;77:327-336
6. Jadhav A. High Strain Rate Properties of Polymer Matrix Composites. Luisiana, EU: Louisiana State University and Agricultural and Mechanical College; 2003
7. Cantwell WJ, Morton J. The impact resistance of composite materials—A review. Composites. 1991;22(5):347-362
8. Barré S, Chotard T, Benzeggagh ML. Comparative study of strain rate effects on mechanical properties of glass fibre-reinforced thermoset matrix composite. Composites Part A: Applied Science and Manufacturing. 1996;27:1169-1181
9. Jacob GC, Starbuck JM, Fellers JF, Simunovic S, Boeman RG. Strain rate effects on the mechanical properties of polymer composite materials. Journal of Applied Polymer Science. 2004;94:296-301
10. Dash K, Sukumaran S, Ray BC. Effect of loading speed on deformation of composite materials: A critical review. Journal of Advanced Research in Manufacturing, Material Science & Metallurgical Engineering. 2014;1:1-22
11. Ploeckl M, Kuhn P, Koerber H. Characterization of unidirectional carbon fiber reinforced polyamide-6 thermoplastic composite under longitudinal compression loading at high strain rate. EPJ Web of Conferences. 2015;94(01041):1-6
12. Koerber H, Camanho PP. High strain rate characterisation of unidirectional carbon-epoxy IM7-8552 in longitudinal compression. Composites Part A:

Applied Science and Manufacturing. 2011;42:462-470

13. Song Z, Wang Z, Ma H, Xuan H. Mechanical behavior and failure mode of woven carbon/epoxy laminate composites under dynamic compressive loading. Composites. Part B, Engineering. 2014;60:531-536
14. Montiel DM, Williams CJ. Method for Evaluating the High Strain-Rate Compressive Properties of Thick Composite Laminates. Research and Development Report, DTRC-SME-90/28. Bethesda, USA: David Taylor Research Center; August 1990
15. Weeks CA, Sun CT. Modeling non-linear rate-dependent behavior in fiber-reinforced composites. Composites Science and Technology. 1998;58:603-611
16. Todo M, Takahashi K, Béguelin P, Kausch HH. Strain-rate dependence of the tensile fracture behaviour of woven-cloth reinforced polyamide composites. Composites Science and Technology. 2000;60:763-771
17. SchoBig M, Bierögel C, Grellmann W, Mecklenburg T. Mechanical behavior of glass-fiber reinforced thermoplastic materials under high strain rates. Polymer Testing. 2008;27:893-900
18. Brown KA, Brooks R, Warrior NA. The static and high strain rate behaviour of a commingled E-glass/polypropylene woven fabric composite. Composites Science and Technology. 2010;70:272-283
19. Jendli Z, Walrick J-C, Bocquet M, Fitoussi J. Strain rate effects on the mechanical behavior of carbon-thermoplastic matrix woven composites. In: ECCM16-16th European Conference on Composite Materials; 2014
20. Walley S, Field J. Strain rate sensitivity of polymers in compression from low to high rates. DYMAT Journal. 1994;1(3):211-227
21. Nemat-Nasser S. Introduction to high strain rate testing. In: ASM Handbook Volume 8 Mechanical Testing and Evaluation. Ohio: ASM International; 2000
22. Qian X, Wang H, Zhang D, Wen G. High strain rate out-of-plane compression properties of aramid fabric reinforced polyamide composite. Polymer Testing. 2016;53:314-322
23. Woo SC, Kim TW. High strain-rate failure in carbon/Kevlar hybrid woven composites via a novel SHPB-AE coupled test. Composites Part B: Engineering. 2016;97:317-328
24. Gray GT III. Classic split-Hopkinson pressure bar testing. In: ASM Handbook Volume 8 Mechanical Testing and Evaluation. Ohio: ASM International; 2000
25. Gray GT III, Blumenthal WR. Split-Hopkinson pressure bar testing of soft materials. In: ASM Handbook Volume 8 Mechanical Testing and Evaluation. Ohio: ASM International; 2000

26. Chen WW, Song B. Split Hopkinson (Kolsky) Bar: Design, Testing and Applications. New York: Springer; 2011

27. Hou JP, Ruiz C. Measurement of the properties of woven CFRP T300/914 at different strain rates. Composites Science and Technology. 2000;60:2829-2834

28. Gama BA, Lopatnikov SL, Gillespie JW Jr. Hopkinson bar experimental technique: A critical review. Applied Mechanics Reviews. 2004;57(4):223-250

29. Reis VL, Marini LF, Donadon MV, Dutra TA, Baldo Junior JE, Nunes de Mello WL. Experimental investigation on the influence of the geometry of carbon fiber specimen used in split Hopkinson pressure bar tests. In: 23rd ABCM International Congress of Mechanical Engineering; 2015

30. Koerber H, Xavier J, Camanho PP, Essa YE, Martín de la Escalera F. High strain rate behaviour of 5-harness-satin weave fabric carbon–epoxy composite under compression and combined compression–shear loading. International Journal of Solids and Structures. 2015;54:172-182

31. Jerabek M, Major Z, Lang RW. Strain determination of polymeric materials using digital image correlation. Polymer Testing. 2010;29:407-416

32. Soutis C. Introduction: Engineering requirements for aerospace composite materials. In: Irving PE, Soutis C, editors. Polymer Composites in the Aerospace Industry. Cambridge: Elsevier Ltd; 2015. pp. 1-18

33. Vodicka R. Thermoplastics for Airframe Applications. A Review of the Properties and Repair Methods for Thermoplastic Composites. Melbourne, Australia: Aeronautical and Maritime Research Laboratory; 1996

34. TenCate. TenCate Cetex and CFRT Thermoplastic Advanced Composites. Brochure. 2016. [Online]. Available: http://www.tencate.com/advancedcomposites/products/thermoplastic/default.aspx

35. Leeser D. Thermoplastic Composites, A Proven Composite Material Technology Generates New Interest. Article. 2010. [Online]. Available: http://www.tencate.com/advancedcomposites/products/thermoplastic/default.aspx

36. Red C. The Outlook for Thermoplastics in Aerospace Composites, 2014-2023: CompositesWorld. Article. 2014. [Online]. Available: http://www.compositesworld.com/articles/the-outlook-for-thermoplastics-in-aerospace-composites-2014-2023 [Accessed: July 25, 2017]

37. ACP Composites. Woven Fabric Style Guide | ACP Composites. 2017. [Online]. Available: https://store.acpsales.com/products/3495/woven-fabric-style-guide [Accessed: August 8, 2017]

38. Mallick PK. Fiber-Reinforced Composites Materials, Manufacturing and Design, Third Editon. Boca Raton: CRC Press Taylor & Francis Group; 2007

39. Silva LFM, Reis VL, Donadon MV, Marques VEC, da Silva VQ, Corat EJ. The strain rate effects on the compressive behavior of composites. In: 22nd International Congress of Mechanical Engineering–Cobem; 2013

40. Greenhalgh ES. Failure Analysis and Fracture of Polymer Composites. Vol. 53(9). Boca Raton: CRC Press LLC; 2009

41. Nemat-Nasser S, Isaacs JB, Starrett JE. Hopkinson techniques for dynamic recovery experiments. Proceedings of the Royal Society of Edinburgh. Section A. Mathematical and Physical Sciences. 1991;435(1894):371-391

42. Field JE, Walley SM, Proud WG, Goldrein HT, Siviour CR. Review of experimental techniques for high rate deformation and shock studies. International Journal of Impact Engineering. 2004;30(7):725-775

43. Reeve S. Introduction to engineering mechanics, analysis, and design. In: ASM Handbook. Volume 21. Composites. Ohio, EU: ASM International; 2001

44. Opelt CV. Estudo dos Modos de Falha em Compressao Uniaxial de Compósitos Avancados: Nova Proposta de Classificacao. São Paulo, Brazil: Instituo Tecnológico de Aeronáutica; 2017

45. Cândido GM, Rezende MC, Donadon MV, De Almeida SFM. Fractografia de compósito estrutural aeronáutico submetido à caracterização de tenacidade à fratura interlaminar em modo I. Polímeros. 2012;22(1):41-53

46. Cândido GM, Rezende MC, Donadon MV, Müller de Almeida SF. Fractografia de comósito estrutural aeronáutico submetido ao ensaio de tenacidade á fratura interlaminar em modo II. Polímeros. 2013;24:18

47. Brent Strong A. High Performance and Engineering Thermoplastic Composites. Pennsylvania: Technomic Publishing Co., Inc; 1993

48. Johnston NJ, Towell TW, Hergenrother PM. Physical and mechanical properties of high-performance thermoplastic and their composites. In: Carlsson LA, editor. Composite Materials Series, 7 Thermoplastic Composite Materials. New York: Elsevier Science Publishers B.V; 1991. pp. 27-71

49. Davies P, Plummer CJG. Structure and mechanical properties of other advanced thermoplastic matrices and their composites. In: Kausch HH, editor. Advanced Thermoplastic Composites: Characterization Processing. Munich: Hanser Publishers; 1993. pp. 141-169

50. dos Santos MN et al. Nanocomposite of photocurable epoxy-acrylate resin and carbon nanotubes: Dynamic-mechanical, thermal and tribological properties. Materials Research. 2013;16(2):367-374

51. Bertholdi J, Opelt CV, Milan JCG, Coelho LAF, Lepienski CM. Propriedades mecânicas, tribológicas e térmicas de nanocompósitos de PLLA com nanotubos de carbono de paredes múltiplas. Polímeros. 2014;24(4):514-520

52. Mouhmid B, Imad A, Benseddiq N, Benmedakhène S, Maazouz A. A study of the mechanical behaviour of a glass fibre reinforced polyamide 6,6: Experimental investigation. Polymer Testing. 2006;25:544-552

53. Abdul Rasheed MI, Rietman B, Visser HA, Akkerman R. Experimental characterisation of recycled (glass/TUP woven fabric) flake reinforced thermoplastic composites. In: The 19th International Conference on Composite Materials; 2017. pp. 3999-4010

3 Chapter

MATRIX CONVERTER FOR AIRCRAFT WITH ENHANCED ELECTRIC SYSTEMS

1. INTRODUCTION: AN OVERVIEW

Power electronics has been crucial in managing electrical drives for industrial and aerospace applications since 1909. However, advancements and innovations in semiconductors have revolutionized the field of power electronics, leading to the creation of numerous converter topologies. There are two types of AC-AC converters that transform a fixed AC voltage and frequency into a variable voltage with variable frequency. Figure 1 displays the configurations of AC-AC converter topologies. [1]:

- DC link
- Direct link

Industries have been utilizing the DC link converter or AC-DC-AC converter since 1902 due to its unique characteristics. For instance, a voltage source inverter provides the following advantages:

- Easy voltage supply control is possible for VSI.
- Low harmonics content exist.

M. Venturini and A. Alesina introduced matrix converter technology in 1981. Their study outlined the basics of matrix converter, including the use of PWM to produce nine pulses with a maximum voltage transfer ratio of 0.5. The primary benefits of MC include high-quality sinusoidal input/output waveforms and built-in regeneration capabilities. In 1986, the authors enhanced the PWM algorithm to achieve a voltage transfer ratio of 0.866 with high-quality sinusoidal output waveforms. Subsequently, numerous articles examined various modulation techniques for MC [4, 5, 7]. The MC struggles with commutating bidirectional switches (BDS); nonetheless, in 1992, four-step commutation [8,9] was implemented. Yaskawa Electric in Japan developed 5.5-kW and 11-kW matrix converters in 2001. Currently, they are working on creating higher-rated matrix converters like the 22 kW and 45 kW versions for lift applications [10] . The Matrix Converter has been evaluated for commercial, industrial [11], and aerospace applications due to its potential benefits[12]. The MC is ideal for aircraft applications due to its ability to offer a wide variety of output frequencies without restrictions, determined by its switching frequency.

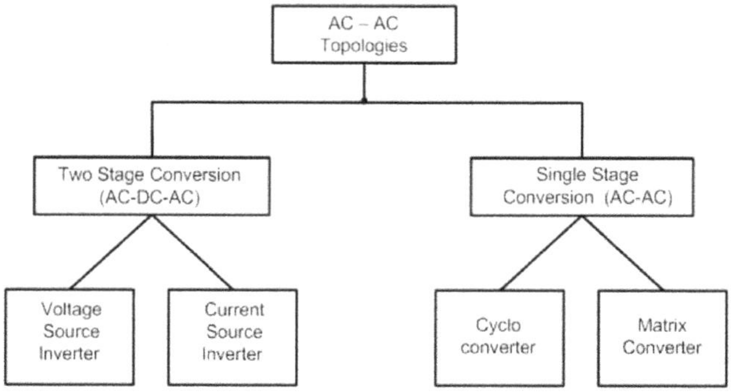

Figure 1. AC-AC power converter topologies.

1.1. Green technology

The goal of More Electric airplanes (MEA) is to promote eco-friendly technology by substituting traditional power sources in airplanes with electrical ones. Conventional aircraft rely on four primary sources of power: electrical, pneumatic, hydraulic, and mechanical. The concept of MEA involves substituting traditional sources of power with electricity generated from environmentally friendly technologies. This chapter discusses green technology for aerospace applications, specifically focusing on aircraft surface actuation control systems. Regenerative power from the MC drive leads to stability issues in the aircraft power supply. It is crucial to address this constraint of the MC drive. The host drum drive motor (HDDM) regenerates power when the tanker aircraft (TA) refueling hose unwinds and winds in the air. Figure 2 displays the circuit of the tanker aircraft with a regeneration control circuit (RCC) designed to dissipate regenerative power of the MC drive using proposed ways.

The host drum drive system (HDDS), which is controlled by Refuelling Control Unit (RCU) and Aeronautical Radio Incorporated Commands (ARINC), controls refueling hose and has three units such as motor control unit (MCU), dump resistor pack (DRP), and two motors. The schematic circuit of HDDS of TA is depicted in Figure 3.

Regeneration occurs only whenever refueling hose winds and trails and this action is commanded by MCU with RCU. It means that the MCU supervises direction of motors based on input from RCU commands. Hence, HDDS must dissipate the regenerated power; otherwise, it can cause below mentioned problems:

- The input supply to HDDS will be increased during regeneration.
- There is possibility of deactivation of HDDS system because of instability in the input supply of HDDS.

The HDDS is using DC link converter, which is not favorable to transient operations of HDDS drive, and this system is bulky and more weight, which are not desirable characteristics for aircraft. The MC drive is proposed to address above problems. However, inherent regeneration capability of matrix converter limits itself being used for above application because bidirectional switches directly fed back to regenerated power to aircraft input power supply without requiring any additional power electronic components.

According to aerospace power quality specifications, this regeneration onto aircraft input power supply must be limited. For this reason, avoidance of regeneration is vital for aircraft surface actuation systems of aircraft. Hence, to avoid regeneration in the matrix converter drive, three novel methods are proposed:

- Bidirectional switch (BDS) method
- Input power clamp (IPC) method
- Standard clamp circuit (SCC) method

To detect regeneration in MC drive, two novel techniques are proposed:

- Power comparison (PC) technique
- Input voltage reference (IVR) technique

Each and every method has its own regeneration control circuit (RCC). For example, RCC for BDS method consists of three bidirectional switches (BDS) in series with three resistors, and this setup is connected across small input filter of MC drive. The RCC for IPC method requires one conventional uncontrolled six-pulse rectifier and a unidirectional switch (UDS) in series with a resistor, and this setup is connected in between input power supply and small input filter of MC. SCC method does not require additional components, which means that no separate RCC is required to do the same action.

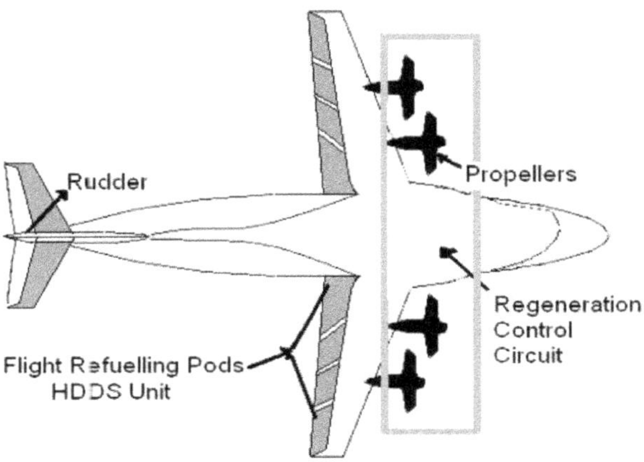

Figure 2. Tanker aircraft with RCC.

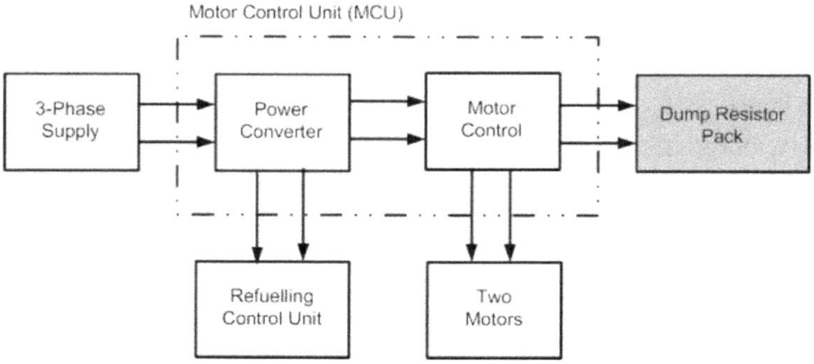

Figure 3. Schematic of host drum drive system of TA.

The simulation results prove that the matrix converter is a suitable alternative to conventional HDDS converter topologies. The BDS method is experimentally adopted to verify the proposed concept by laboratory prototype matrix converter, which is built at Smiths Aerospace laboratory (later called GE Aviation laboratory) in University of Nottingham. This MEA project strongly supports the green environment by adopting abovementioned green technologies to obtain reduced aircraft emissions.

2. MATRIX CONVERTER MODULATION AND VECTOR CONTROL

Matrix converter consists of nine bidirectional switches arranged in 3 × 3 as shown in Figure 4. This is called all silicon solution [13]. The input phases (A, B, C) can be connected to output phases (a, b, c) for any switching period of time using bidirectional switches. The switches are controlled in such a way that the average output voltage is a sinusoidal waveform of the desired frequency and amplitude.

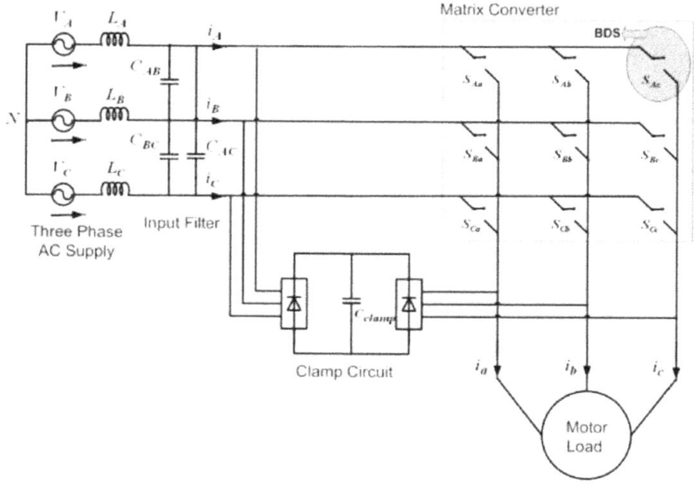

Figure 4. The general structure of the conventional MC.

2.1. Basic Rules

The matrix converter consists of nine bidirectional switches with 29 (512) possible switching states. However, only 27 switching states can be used because two basic rules have to be followed [13].

- No short circuit of two inputs
- Never open circuit the outputs

Because of the above rules and also inductive loads nature of MC drive, each output line must always be connected to an input line. Under these basic rules, space vector modulation (SVM) for MC drive has 27 switching states.

2.2. Space Vector Modulation

The space vector modulation (SVM) is defined as type of pulse width modulation (PWM) to generate gate drive signal to trigger the bidirectional switches (BDS) in MC [14]. SVM is also preferable to control and analyze machines with vector control (VC) or field oriented control of machines and allows visualization of the spatial and time relationships between the resultant current and flux vectors (or space phasors) in various reference frames.

2.3. Vector Control

Decoupling flux and torque is feature of VC to overcome sluggish torque response of Induction Motor (IM) to work like a separately excited DC machine. To achieve an independent control of the flux and torque, the direct axis (*d* axis) is aligned to a rotor flux vector (*r*) and the concept of the indirect field-oriented vector control (IFOVC) is depicted in Figure 16 ,15] 5, 17].

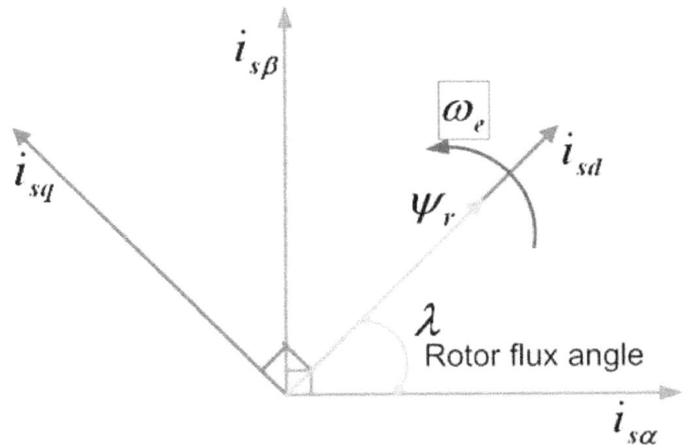

Figure 5.Field orientation: Ψ_r is aligned with d-axis.

The rotating reference frame is rotating at synchronous angular velocity (ω_e). The sensed three-phase output currents of MC drive are converted into stationary reference frame (i_s, i_s) and then viewed as two "dc" quantities (i_{sd}, i_{sq}). The direct axis or real axis component is responsible for the field producing current (i_{sd}) and is ideally maintained constant up to the motor synchronous speed. If d-axis is aligned with rotor flux vector ($_r$), the system is said to be field oriented. The q-axis component is responsible for torque producing current (i_{sq}). These two vectors are orthogonal to each other so that the field current and torque current can be controlled independently [16, 17].

2.4. Closed-Loop IFOVC

Both faster current control loop and speed control loop outputs [12] provide the reference voltages (V_{sd}, V_{sq}) and hence (V_a, V_b, V_c) to SVM to get the stable operation of MC drive as shown in **Figure 6.**

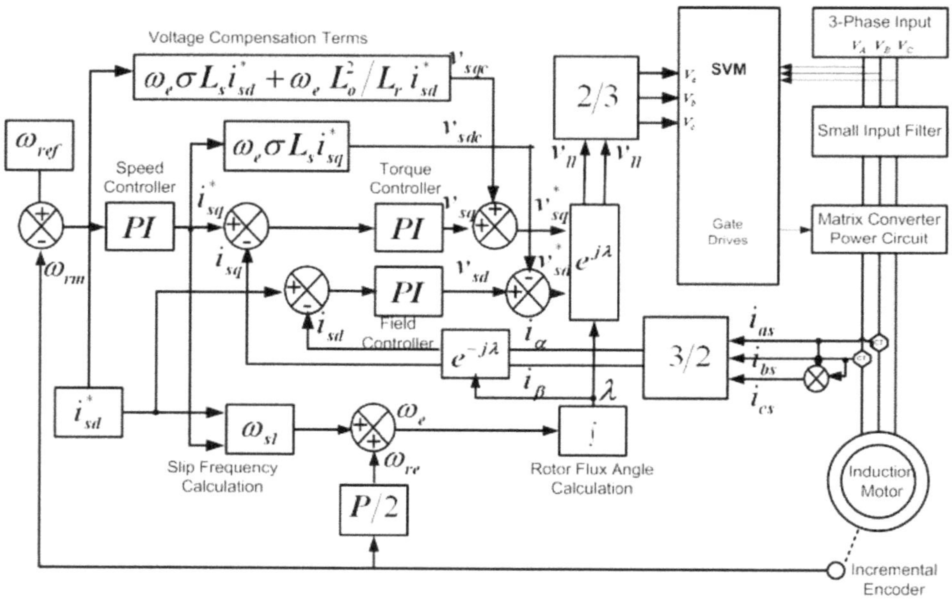

Figure 6. Closed-loop indirect field-oriented vector control (IFOVC) scheme.

3. REGENERATION DETECTING TECHNIQUES

Two novel techniques [18] are used to identify regeneration when step is applied to reverse the MC drive. These are (1) power comparison technique (PC) and (2) input voltage reference (IVR) technique. These techniques are responsible for generating pulses for regeneration control circuit (RCC) or electrical braking circuit (EBC) whenever regeneration is detected in the MC drive. In PC, output power is used as reference; hence, it is called PC technique similarly in IVR technique, the voltage across the small input filter capacitor and output power both are used as reference; hence, it is called IVR technique. The IVR technique is similar to and derived from conventional dynamic braking technique.

3.1. Power Comparison (PC) Technique

To calculate the absolute value of output power of MC drive to achieve power comparison (PC) technique, the torque producing current (i^*_{sq}) and measured rotor speed (ω_{re}) are sensed. Figure 7 shows the gate drive signal, which is generated for RCC of input power clamp (IPC) method. Here power dissipation through a resistor in the regeneration control circuit (RCC) is directly proportional to the duty cycle of unidirectional switch (UDS), as in Eq. (1),

$$P_{dis} \propto D \tag{1}$$

where D = duty cycle of the unidirectional switch and P_{dis} = power dissipation through the resistor.

The duty cycle calculation requires the maximum electrical braking power (P_{mb}) to be calculated, as shown in Eq. (2). The duty cycle of the switches is then less than or equal to unity under all operating conditions.

$$P_{mb} = T_{me}\,\omega_{mre} \tag{2}$$

where T_{me} = electromagnetic torque and ω_{mre} = speed of MC drive.

The gate drive signals for RCC switches are generated by using field programmable gate array (FPGA) with digital signal processor (DSP). Here FPGA that receives input parameters (ω_{re}, T_e, i^*_{sq}) from sensors is fed into DSP, which does all mathematical calculations to generate gate drive signal as shown in Figure 7, and again fed back to FPGA that is sending gate drive signal to the gate drive of UDS. The duty of UDS/BDS is linearly varying with respect to output negative power.

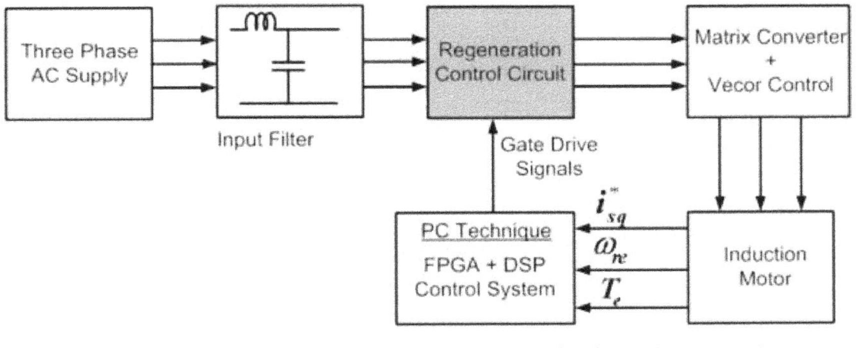

Figure 7. Block diagram of the power comparison (PC) technique for IPC method.

The MC drive is not capable to output whole of regenerated power because of it losses such as friction, windage, iron, switching and conduction losses. Because of above reason, the braking resistor dissipates less than the actually regenerated power. As written in Eqs. (3) and (4), the design of braking resistor (R_b) relies on maximum regenerative power during regeneration.

$$P_{in,max} = V_{in}^2/R_b \quad (3)$$

$$I_b = V_{in}/R_b \quad (4)$$

where the braking current (I_b) and input power ($P_{in,max}$) are directly proportional to the input voltage. The braking resistor design also depends on the braking time, thermal capacity of the resistor, and heat sink. And the current rating of UDS/BDS in RCC must be higher than the braking current.

3.2. Input Voltage Reference (IVR) Technique

The voltage across small input filter capacitor is measured and compared to the MC supply voltage to generate gate drive signal for RCC of IPC method as shown in Figure 8.

The IVR technique can be used to detect the regeneration in the matrix converter for electrical braking methods. The duty cycle variation is directly proportional to the increase in the line to line voltage across the input filter capacitor of the matrix converter under regeneration with respect to the output power (P_o), as shown in Eq. (5).

$$V_{AB} \propto P_o \propto D \quad (5)$$

Here, V_{AB} is line voltage across small input filter capacitor.

The RCC is turned on only if any voltage difference is detected between MC supply voltage and voltage across the small input filter capacitor for each sampling period. In addition, if the small input filter capacitor voltage is equal to the MC supply voltage, then the duty cycle is set to zero. Gate drive signal is generated using FPGA and DSP control platform similar to PC technique. The power dissipation through RCC happens only if the MC drive is operating under regenerative mode.

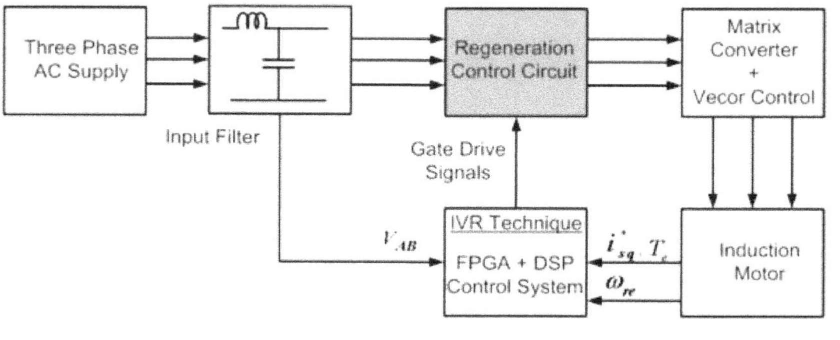

Figure 8. Block diagram of the input voltage reference (IVR) technique for IPC method.

4. METHODS FOR AVOIDING REGENERATION IN MATRIX CONVERTER

To avoid regeneration in matrix converters, three novel circuit topologies are investigated:

- Bidirectional switch (BDS) method
- Input power clamp (IPC) method
- Standard clamp circuit (SCC) method

4.1. Bidirectional Switch (BDS) Method

The power circuit for the BDS method [18], regeneration control circuit (RCC) or electrical braking circuit, is shown in Figure 9. The regeneration control circuit (RCC) is introduced across the input filter capacitors (C_{AB}, C_{BC}, and C_{AC}). The regeneration control circuit (RCC) is responsible for power dissipation when regeneration takes place in the MC motor drive.

The RCC consists of three bidirectional switches (BDS_{AB}, BDS_{BC}, and BDS_{AC}) in series with three resistors (R_{AB}, R_{BC}, and R_{AC}) connected across the input lines, in parallel with the input filter capacitor. The schematic of the regeneration control circuit (RCC) or electrical braking circuit (EPC) is depicted in Figure 9.

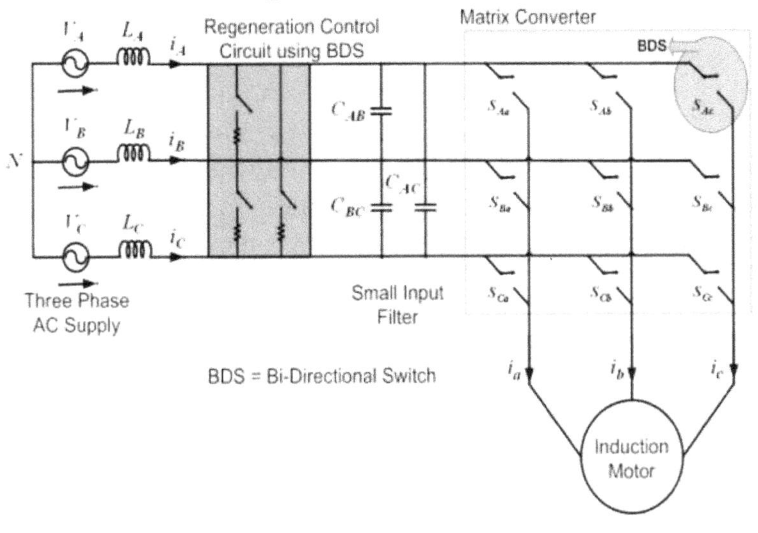

Figure 9. Bidirectional switch (BDS) method.

4.2. Input Power Clamp (IPC) Method

The input power clamp (IPC) method [19] is used for braking the electrical energy in a matrix converter motor drive. The IPC method requires only one braking resistor and a UDS, as shown in Figure 10, when compared to the BDS method, which requires three switches in series with three resistors. Electrical braking circuit or regeneration control circuit (RCC) for the input power clamp (IPC) method is located across the input filter capacitors (C_{AB}, C_{BC}, and C_{AC}). The RCC of IPC is controlled using either power comparison (PC) technique or the input voltage reference (IVR) technique.

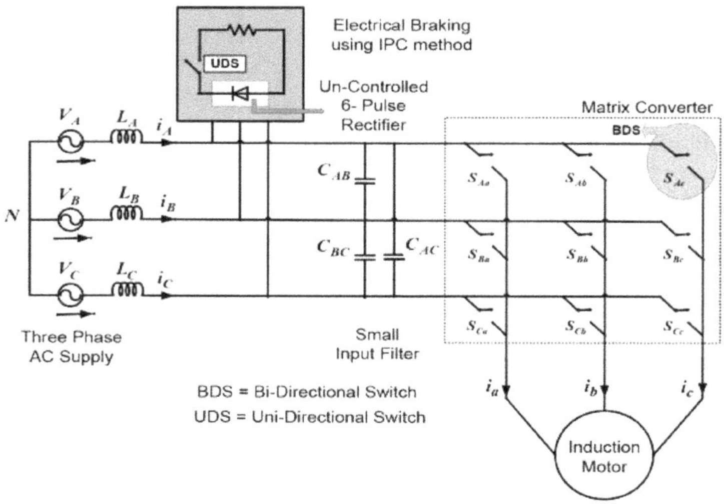

Figure 10. The input power clamp (IPC) method.

The main power electronic components for RCC of IPC are conventional uncontrolled six-pulse rectifier and a UDS in series with a braking resistor (R), as shown in Figure 10. This braking resistor does not have inductive property to help to achieve better electrical braking when regeneration happens in MC drive. It is believed that IPC method is the best method when compared to BDS method because it requires only fewer power semiconductor switching components but not suitable for aerospace applications because it has electrolytic capacitor in the RCC.

4.3. Standard Clamp Circuit (SCC) Method

Similar to the BDS method and the IPC method [19], the proposed standard clamp circuit (SCC) method is using two techniques to detect the regeneration in the matrix converter drive. These are (1) power comparison (PC) technique and (2) input voltage reference technique. However, here the PC technique is only considered and the simulation results of the SCC method with PC technique are discussed. The block diagram for standard clamp circuit is shown in **Figure 11.** The electrical braking circuit for SCC is shown in **Figure 14.**

Power dissipation through resistor is directly proportional to duty cycle of UDS, which is already given in Eq. (1). To prove the performance of the SCC method for electrical braking in the MC drive, a 2.2-kW vector-controlled induction motor fed by MC drive is considered.

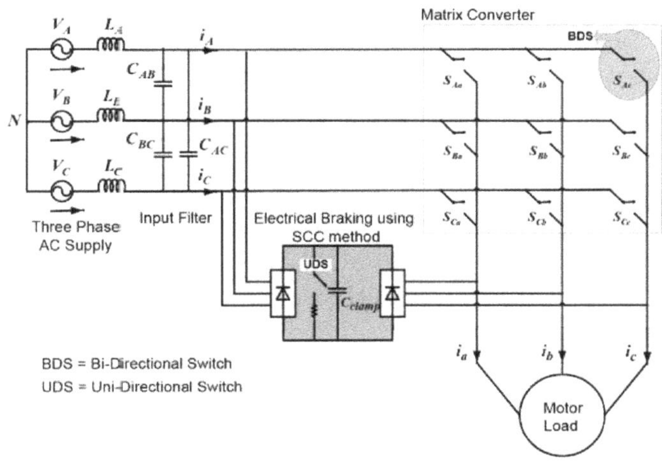

Figure 11. The standard clamp circuit method.

4.4. Comparison of BDS, IPC, and SCC Methods

When compared to earlier methods, called the BDS method and IPC method, no auxiliary hardware is required for SCC method for electrical braking in the matrix converter drive as shown in Table 1. The BDS method [18] has three drawbacks:

1. It requires three BDS in series with three resistors.
2. Also, it requires complex control platform, which controls six PWM for electrical braking.
3. This auxiliary circuit increases size, weight, and cost.

Similarly, the IPC method has three main drawbacks:

1. It requires conventional uncontrolled diode rectifier and UDS with a resistor.
2. Separate control platform for a PWM for electrical braking is required.
3. This auxiliary circuit increases the size, weight, and cost.

The SCC method requires only one UDS switch in series with a resistor. Because of using SCC in the MC drive, achieving electrical braking using this method is easy and no complicated control platform is required. The SCC is considered as a safety device to protect the matrix converter under abnormal conditions such as overvoltage in the input side or output side.

Table 1. Comparison of BDS, IPC, and SCC.

Factors	BDS method	IPC method	SCC method
Switches	3 (BDS)	1 (UDS)	1 (UDS)
Resistors	3	1	1
Diodes	0	6	0
Weight, size, and cost	Considerably increased	Bit increased	Remains same
Implementation	Complicated	Not complicated	Simple

5. SIMULATIONS RESULTS

A simulation analysis is conducted using SABER software package to forecast and confirm the effectiveness of the proposed techniques for preventing regeneration in a matrix converter. [12].

5.1. Regeneration

The regeneration can be demonstrated at V_{in} = 240 V, q = 0.75, and f_s = 10 kHz by applying step transient to reverse the speed at 1.2 s as shown in **Figure 12.** A step transient lasts upto 2.4s, no load speed reversal (from +188.5 rad/s to -188.5 rad/s), as shown in the **Figure 13** which also shows developed torque which is direclty proportional to torque producing current (i_{sq}) of IM during VC. The torque producing current (i_q) of the induction motor reaches the maximum limit of 35 A during acceleration as shown in **Figure 13.** Here regenerative power depends upon large inertial load (j = 0.089 kg m^2) of the induction motor, which is created coupling IM with the same rating of DC motor (4 kW). The output current waveforms of the matrix converter drive is depicted in **Figure 14(a),** which also indicates during speed reversal, the four-quadrant operation, inherent property, of MC from motoring mode to regenerating mode is smoothly achieved. The control of dq-currents (i_d, i_q) with no coupling effects is demonstrated. **Figure 14(b)** shows input phase currents (i_A, i_B, i_C) of the matrix converter during the four-quadrant operation. The input regenerative powers (P_A, P_B, P_C) to be dissipated using the regeneration control circuit (RCC) are shown in **Figure 15(a).**

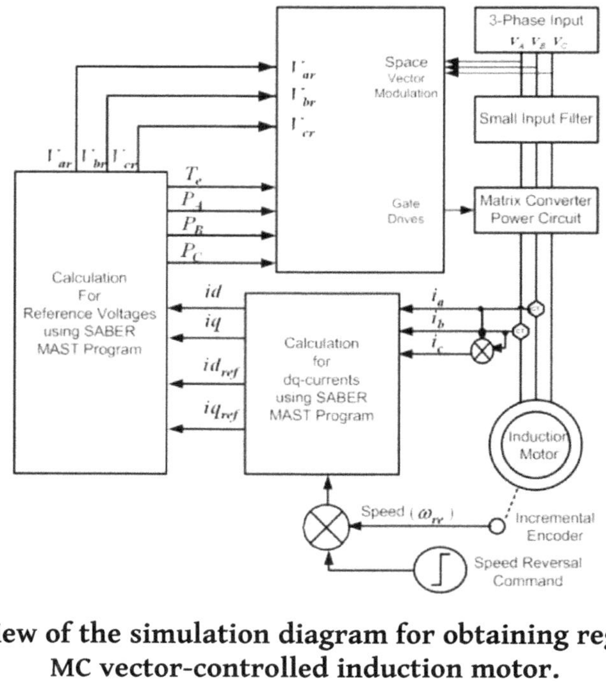

Figure 12. Overview of the simulation diagram for obtaining regeneration in the MC vector-controlled induction motor.

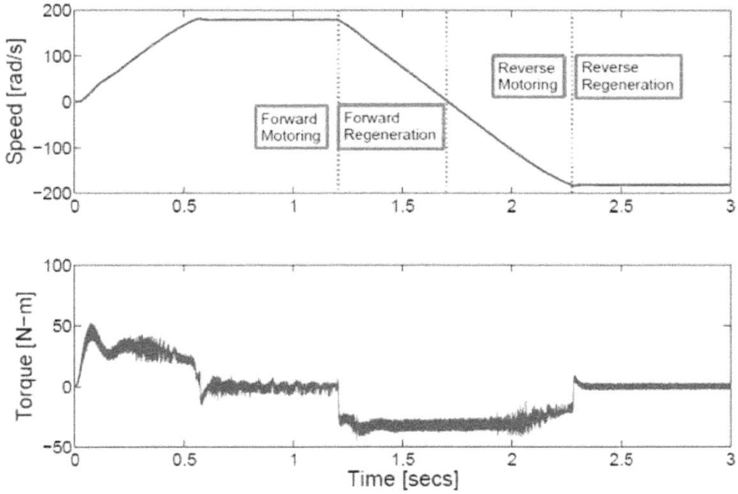

Figure 13. Speed and torque of the vector-controlled IM in regeneration. V_{in} = 240 V, q = 0.75, and f_s = 10 kHz.

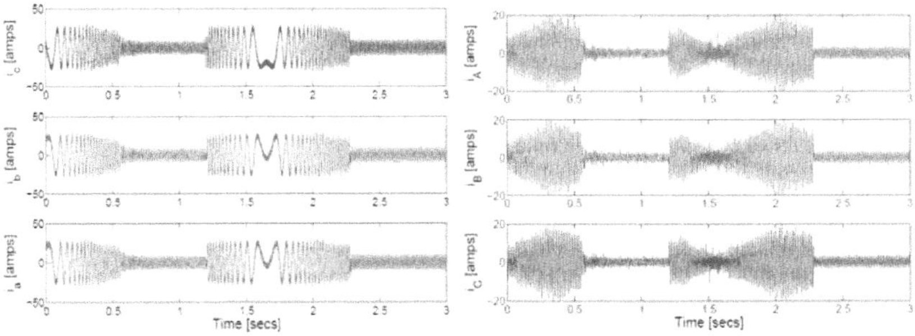

a) b)

Figure 14. (a) Output currents of the MC and (b) input phase currents of the MC.

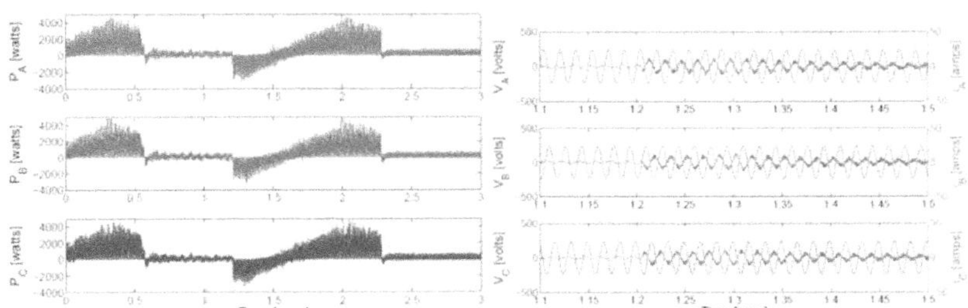

a) b)

Figure 15. (a) Input phase powers of the MC and (b) phase opposition at regeneration.

During regeneration, the phase opposition (180° phase displacement) between the input phase voltages (V_A, V_B, V_C) and input phase currents (i_A, i_B, i_C) can be seen in **Figure 15(b)**.

5.2. BDS Method with PC Technique

The generation of the required identical six pulses using PC technique for the regeneration control circuit (RCC) of BDS method is shown in **Figure 16(a)**. Here duty cycle is linearly varying with respect to the output power of the MC drive as shown in **Figure 16(b)**. In order to verify the regenerative energy dissipation, the input phase powers are calculated using input phase voltages and the input phase currents. The resulting input phase currents and calculated input phase power are shown in **Figure 17(a)** and (b), respectively. When compared to **Figure 15(b)**, **Figure 18** proves that regenerative power is dissipated using novel RCC of BDS method, hence input phase voltages (V_A, V_B, V_C) and input phase currents (i_A, i_B, i_C) are in phase. However, there is some input power left, as shown in **Figure 17(b)**,because of constant losses (such as friction, windage losses, and inertial losses) in the IM and switching noises.

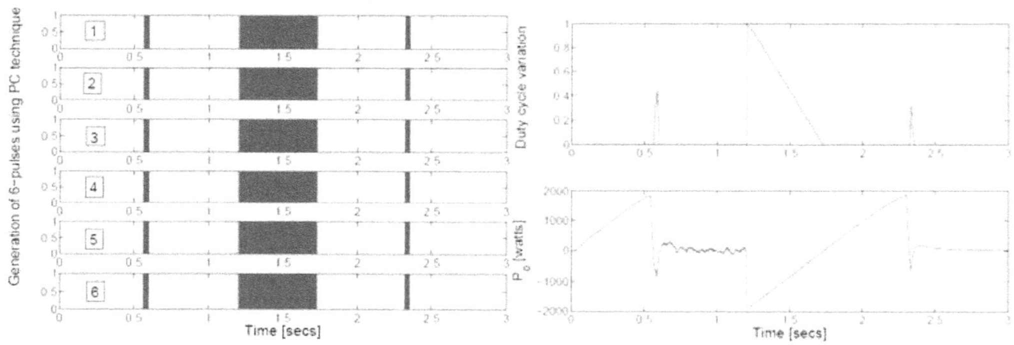

a) b)

Figure 16. (a) Generation of the six pulses and (b) duty cycle and output power variation for regeneration control circuit of the BDS method with the PC technique. V_{in} = 240 V, q = 0.75, and f_s = 10 kHz.

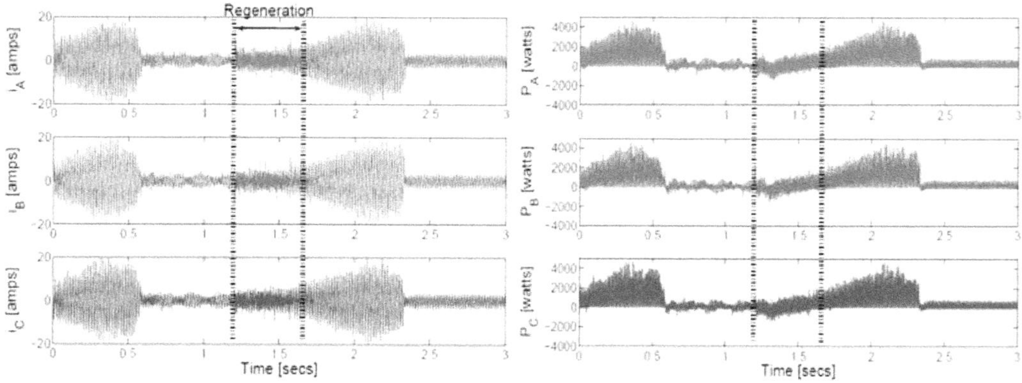

a) b)

Figure 17. (a) Input phase currents and (b) input phase powers for BDS method.

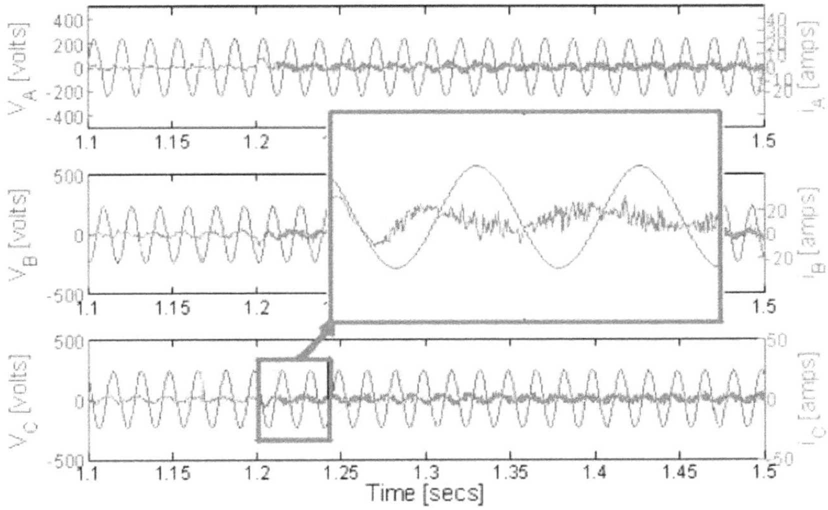

Figure 18. Phase relationship between input phase voltages and currents of MC drive.

5.3. IPC Method and SCC Method with IVR Technique

The generation of a pulse to trigger the RCC, duty cycle variation, and the output power (P_o) variation during regeneration for the IPC method with the IVR technique is shown in Figure 19. Figure 20 shows input phase powers (P_A, P_B, P_C) of MC drive after regeneration control. From simulation results, the IVR technique for both methods (IPC and SCC) is producing acceptable results to avoid regeneration with a MC drive similar to PC technique.

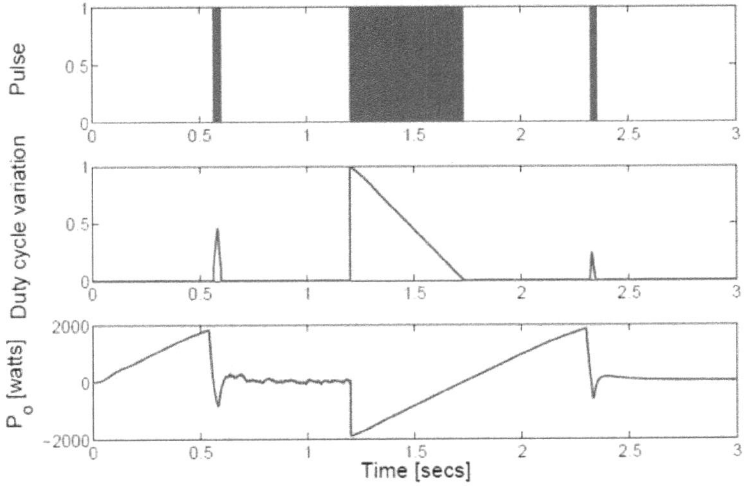

Figure 19. Pulse for RCC, duty cycle variation, and output power for the IPC method with the IVR technique. V_{in} = 240 V, q = 0.75, and f_s = 10 kHz.

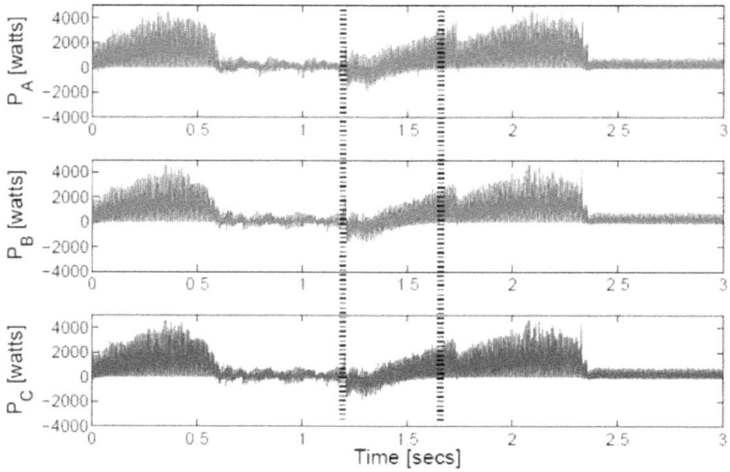

Figure 20. Input phase powers for the SCC method with the IVR technique. V_{in} = 240 V, q = 0.75, and f_s = 10 kHz.

6. EXPERIMENTAL ANALYSIS

The control platform includes both the control circuits and the interface circuits. The control circuit consists of a DSP card and an FPGA card as shown in Figure 21. The interface circuit includes encoder interface board and DSP daughter card (C6713DSK HPI), which is used to send user inputs and output waveforms plotting to troubleshoot hardware problems that are faced during hardware implementation and achieving desired output results. For example, if spike occurs while reversing the speed of the MC drive, the error data were captured and troubleshot through DSP daughter card. Figure 22(a) shows the RCC of three bidirectional switches (BDS_{AB}, BDS_{BC}, BDS_{AC}) that are connected between MC input supply line voltages (V_{AB}, V_{BC}, V_{AC}) and the small input filter capacitors (2).

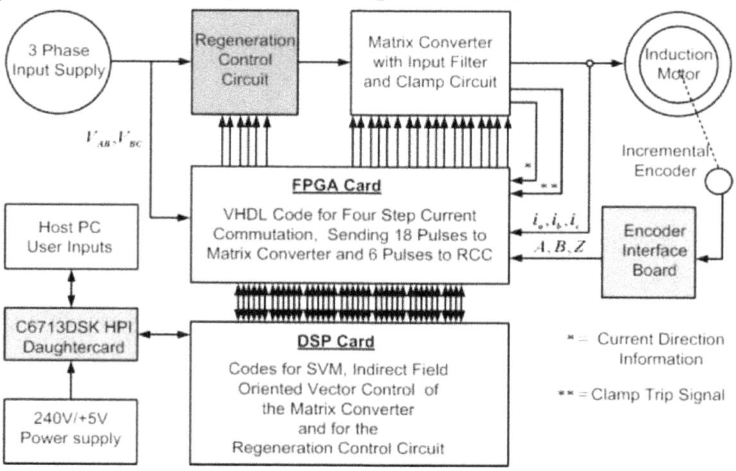

Figure 21. Layout of control and interface circuits.

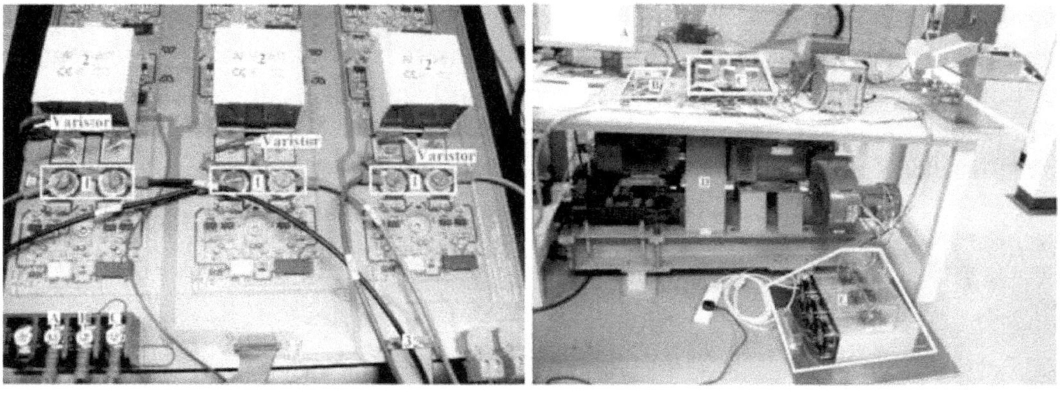

a) b)

Figure 22.(a) Photograph of RCC for BDS method and (b) complete experimental setup.

The RCC resistors (R_{AB}, R_{BC}, R_{AC}) are connected in series (1) with the bidirectional switches. The triggering pulses (3) for bidirectional switches are obtained from FPGA card. Figure 22(b) shows complete experimental setup of MC drive. The host user interface PC is used to help monitor inputs to the system (complete experimental setup of MC drive) and get output from the system. The power circuit and control circuit of the laboratory prototype matrix converter drive is highlighted by letters (C) and (B), respectively. Letters (D) and (E) indicate the 4 kW induction motor with high inertial load and the RCC resistors with their heat sink arrangement, respectively.

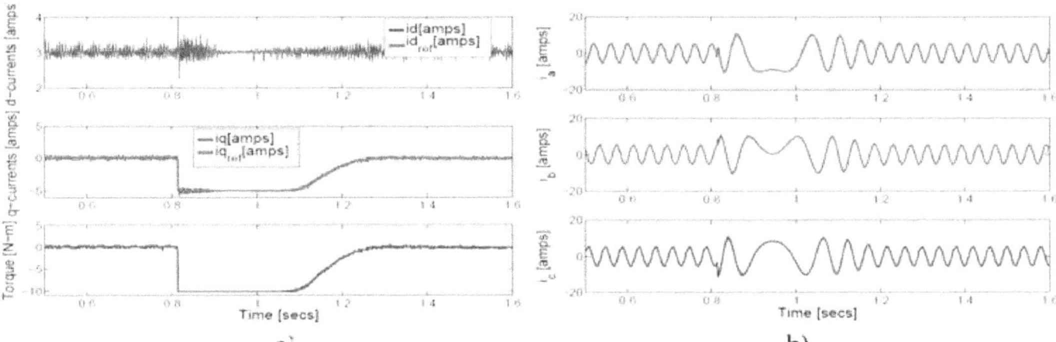

a) b)

Figure 23.(a) Torque, dq-currents and (b) stator currents of the motor.

Proof of BDS method for electrical braking in the MC drive by carried out experiments, at Smiths aerospace (later called GE Aviation) laboratory in PEMC Group of University of Nottingham, using a prototype rated at 7.5 kW MC fed a 4 kW IM. The field current (d-currents), torque current (q-currents), torque of IM and stator currents of IM during regeneration at speed reversal from +157 rad/s to -157 rad/s [V_{in} = 200 V, f_s = 12.5 kHz, q = 0.75] are shown in Figure 23(a) and Figure

23 (b), respectively. The input phase voltages (V_A, V_B), input phase currents (i_A, i_B), and three phase input power (using two-wattmeter method) during regeneration and after avoiding regeneration are shown in Figure 24(a) and (b), respectively. The above experimental results (from Figure 24(a) and (b)) clearly show that the regenerative (negative) power is dissipated through the RCC.

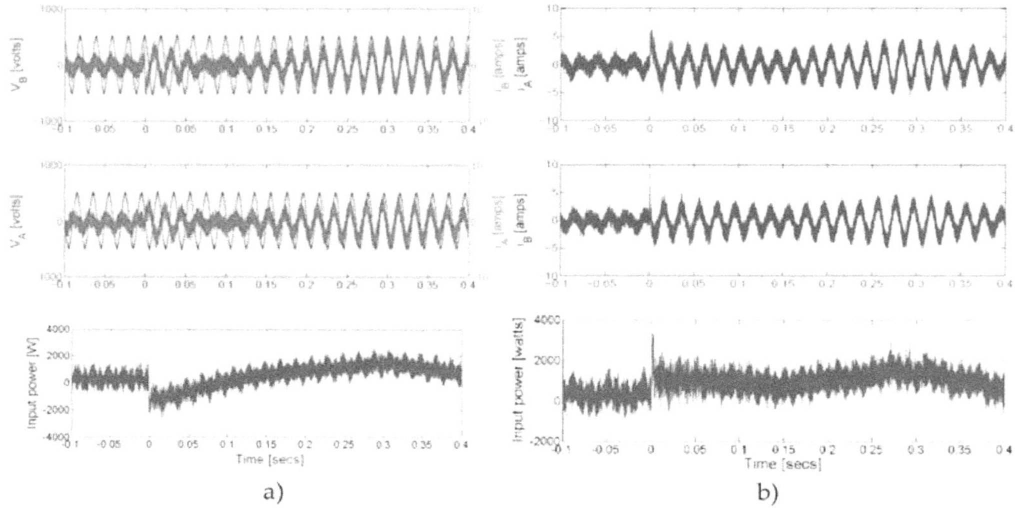

Figure 24.(a) During regeneration and (b) after avoiding regeneration.

7. CONCISE SUMMARY

Matrix converter (MC) technology has been favored over other direct AC/AC converters or AC-DC-AC link converters since 1989 due to its unique characteristics, including the absence of DC link components, high-quality sinusoidal input/output waveforms, inherent regeneration capability, and unrestricted output frequency. The published research on the matrix converter centers around the following concepts:

- MC for aerospace and industrial applications
- Power quality and stability of MC
- Addressing commutation techniques to avoid failure of the power circuit of MC
- Modulation topologies for the MC

This research is focused on the application of a matrix converter for more electric aircraft (MEA). To make MC appropriate for aerospace applications, the inherent regeneration must be avoided, which eliminates the distinctive attribute of four-quadrant functioning of MC, in order to meet the aircraft power quality criteria. Prior to this study, there was a lack of focus on this research field, which will enable the utilization of the matrix converter in aerospace and certain industrial settings. For example, at the beginning of the twenty-first century, the matrix converter has been made as a commercial product, lift, which is manufactured by Yaskawa (Japan). The PEMC Group at the University of Nottingham is working on creating a 150-kVA matrix converter for more powerful uses. The typical clamp circuit method with power comparison technique is preferred due to its lack of need for additional hardware, despite the good results that can be achieved using BDS, IPC, and SCC methods. Therefore, the matrix converter's weight, size, and cost are significantly decreased. Thus, the matrix converter with SCC technique is suggested for aerospace applications where regeneration into the supply is prohibited. Based on the experimental results, it is determined that using electrical brakes with a matrix converter motor is viable for aerospace applications such more electric aircraft.

8. AT A GLANCE

This chapter proposes three approaches that prevent regenerative power from the matrix converter (MC) motor drive to be transferred to the aircraft power supply. As per aerospace power quality standards, regenerative power must to be absorbed within the drive to prevent instability issues in the aircraft's power system. Three methods used for aerospace applications include bidirectional switch (BDS), input power clamp (IPC), and standard clamp circuit (SCC). Two innovative strategies are suggested to detect regeneration in a matrix converter drive. These techniques are power comparison (PC) and input voltage reference (IVR). Both strategies utilize the output power of the motor controller and the speed direction to identify regeneration in the motor controller drive. Electrical braking is crucial in several aircraft applications like surface actuation and air-to-air refueling systems. Hence, the matrix converter drive's intrinsic regeneration capacity is unsuitable for aircraft applications and should be prevented. The proposed methodologies are illustrated through thorough simulation outcomes and experimental confirmation. A 7.5-kW matrix converter was used to power a 4-kW induction motor with inertial load in an experimental setup to validate new procedures. The results acquired using the BDS approach with PC technology demonstrated the feasibility of avoiding regeneration with a matrix converter. This chapter is significant for the 150-kVA matrix converter in high-power applications.

REFERENCES

1. Bose BK. Energy, environment and advances in power electronics. IEEE Transactions on Power Electronics. 2000;15:688-701. DOI: 10.1109/63.849039

2. Lipo TA. Recent progress in the development in solid-state ac motor drives. IEEE Transactions on Power Electronics. 1998;3:105-117. DOI: 10.1109/63.4340

3. Gyugyi L, Pelly B. Static Power Frequency Changers: Theory, Performance and Applications. New York: John Wiley and Sons; 1976
4. Alesina A, Venturini M. Solid-state power conversion: A fourier analysis approach to generalized transformer synthesis. IEEE Transactions on Circuit and Systems. 1981;28:319-330. DOI: 10.1109/TCS.1981.1084993
5. Alesina A, Venturini MGB. Analysis and design of optimum amplitude nine-switch direct ac-ac converters. IEEE Transactions on Power Electronics. 1989;4:1284-1291. DOI: 10.1109/63.21879
6. Maamoun A. Development of cycloconverters. In: Proceedings of the IEEE Canadian Conference on Electrical and Computer Engineering. Canada: IEEE; 2003. pp. 521-524
7. Helle L, Larsen KB, Jorgensen AH, Munk-Nielsen S, Blaabjerg F. Evaluation of modulation schemes for three-phase to three-phase matrix converters. IEEE Transactions on Industrial Electronics. 2004;51:158-171. DOI: 10.1109/TIE.2003.821900
8. Wheeler PW, Grant DA. A low loss matrix converter for variable speed drives. In: Proceedings of the IEE Conference (IEE'93). UK: IEE; 1993. pp. 27-32
9. Burany N. Safe control of four quadrant switches. In: Proceedings of the IEEE Industrial Applications Society (IAS'89). USA: IEEE; 1989. pp. 1190-1194
10. Yaskawa Electric Corporation. Matrix converter drive varispeed ac. In: Inverter Drive on Yaskawa Technical Review. Japan: Yashkawa Electric Corporation; 2007
11. Klumpner C, Nielsen P, Boldea I, Blaabjerg F. A matrix conveter motor(mcm) for industry applications. IEEE Transactions on Industrial Electronics. 2002;49:325-335. DOI: 10.1109/41.993265
12. Imayavaramban M. Avoiding regeneration with a matrix converter drive [PhD thesis]. UK: The University of Nottingham; 2009
13. Wheeler PW, Rodriguez J, Clare JC, Empringham L. Matrix converter—A technology review. IEEE Transactions on Industrial Electronics. 2002;49:276-288. DOI: 10.1109/41.993260
14. Huber L, Borojevic D. Space vector modulator for forced commutated cycloconverters. In: Proceedings of the IEEE Industrial Applications Society (IAS'89). USA: IEEE; 1989. pp. 871-876
15. Krishnan R, Bharadwaj AS. A review of parmeter sensitivity and adaptation in indirect vector controlled induction motor drive systems. In: Proceedings of the IEEE Power Electronics Specialists Conference (PESC'90). USA: IEEE; 1990. pp. 560-566

16. Blaschke F. The principle of field orientation as applied to the new transvector closed loop control for rotating machines. Seimens Review. 1972;34:217-220

17. Leonhard W. Control of Electrical Drives. Berlin: Springer-Verlag; 1985

18. Imayavaramban M, Wheeler PW. Avoiding regeneration with a matrix converter drive. In: Proceedings of the IEEE Power Electronics Specialists Conference (PESC'07); June 2007; Florida. USA: IEEE; 2007. pp. 2529-2534

19. Imayavaramban M, Wheeler PW, Empringham L, Clare JC. Input power clamp for controlling regeneration in matrix converter drive. In: Proceedings of the IET Inernational Conference on Power Electronics and Machines Drives (PEMD'08); April 2008; York. UK: IET; 2008. pp. 515-519

Chapter 4

MECHANISMS OF DEGRADATION IN TURBINE ENGINE LUBRICANTS AND ADDITIVES

1. INTRODUCTION: AN OVERVIEW

Lubrication is crucial in situations that include the movement of components. Aircraft propulsion systems consist of several moving components that operate at high speeds in extreme temperatures and conditions. Turbine engine lubricants are crucial for minimizing wear, decreasing friction, and dispersing heat generated by the engine. Contemporary engines are engineered to function at elevated temperatures and shear rates, which require lubricants and additives to meet greater performance standards. Turbine engine lubricants usually contain a basestock composed of synthetic esters and other additives that alter the basestock's characteristics. Additives are added to the basestock to decrease oxidation, minimize metal bearing wear, or alter lubricant characteristics. [1].

When exposed to high temperatures, all lubricants degrade, leading to alterations in both their physical and chemical characteristics. Physical property changes may involve alterations in viscosity, variations in boiling point, or shifts in freezing point, among other factors. Chemical qualities that are subject to change are corrosion of metals, polymer formation, and oxidation of the base stock. Lubricants contain additives that alter the characteristics of the basestock. Deterioration of the additives diminishes their efficacy and may lead to engine malfunction. Within turbine engine operations, the reduction of additives is a crucial indicator of the condition of the lubricant. Lubricant deterioration is crucial for engine health and has substantial implications for human health. Most commercial airplanes obtain cabin pressurization air from the engine immediately after the compressor section. Degradation byproducts of lubricants have been demonstrated to migrate from the engine to the cabin when seals fail, causing significant health consequences. The usual low level loss of lubricants and degradation products into the cabin under normal flying conditions is of significant importance. Seals are known to leak, and part of the released material can enter the passenger cabin in the form of fumes and nano-droplets. The long-term harmful effects of these substances are a significant worry [2].

This chapter will present the composition of typical turbine engine lubricants in Section 2. Section 3 discusses the disintegration methods of the basestock, followed by the additive degradation mechanism in Section 4. Section 5 analyzes synergistic and antisynergistic interactions between lubricants and additives. The document will cover modifications in bearing systems, the integration of ionic liquids and nanoparticles, and will conclude with an analysis of the effects of lubricant deterioration in Section 6.

2. COMPOSITION OF TURBINE ENGINE LUBRICANTS

Turbine engine lubricants have changed dramatically over the years in response to the increasing stresses applied to the lubricant. In particular higher shear stress, higher operating temperatures and lower storage temperatures have made changes in both basestocks and additive packages necessary. Natural petroleum based oils could not meet the temperature demands which made the selection of synthetic materials, modified with a number of additives necessary for this application [3]. In order to meet the demands for modern aircraft, lubricants based on synthetic esters were developed and have been refined many times, both in terms of the basestocks and the additive packages to meet the current specifications.

2.1. Basestock Composition

The composition of lubricant basestocks for turbine engines is somewhat variable as long as they can meet the performance requirements set forth in the standards SAE5780 for commercial aircraft and either MIL-PRF 23699 [4] or MIL-PRF 7808 [5] for military aircraft. One of the requirements is to be compatible with all of the previously approved lubricants in a given specification to avoid the inevitable mixing. Esters have been used since the 1940 as synthetic basestocks that have desirable thermal properties, however no single ester meets all requirements. Modern lubricant basestocks use a mixture of a number of esters in order to tailor the properties of the lubricant to the desired properties. These specifications have resulted in the use of certain common ester basestocks. Ester basestocks for turbine engines are all ester based using polyols and common carboxylic acids. Some of the common alcohols used are shown n Figure 1.

The polyols shown have been selected because they are highly hindered and also lack hydrogen atoms in the position. Previous studies have shown that increases thermal and hydrolytic stability results when there is no hydrogen atom present on the carbon atom. The carboxylic acids used to make the esters are a combination of linear and branched acids with a blend being frequently used to arrive at the desire viscosity. Normally 5 cs (SAE5780 and MIL-PRF 23699) basestocks use pentaerythritol and dipentaerythritol for the alcohols and C5-C10 linear and branched acids. Lower viscosity lubricants (Mil-PRF 7808) are of the based on neopentyl glycol and trimethylolopropane as the alcohol and C5-C12 linear carboxylic acids. The incorporation of branched acids in lubricants has a significant effect on the thermal stability and physical properties of the lubricant. Some of the different acids are shown in Figure 2.

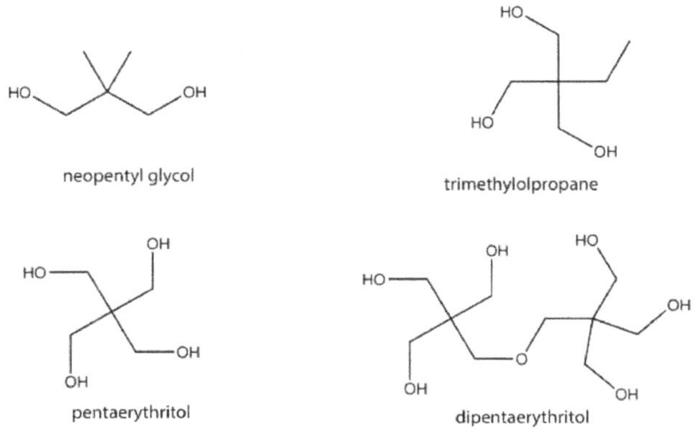

Figure 1. Common polyols used to make ester-based lubricant basestocks.

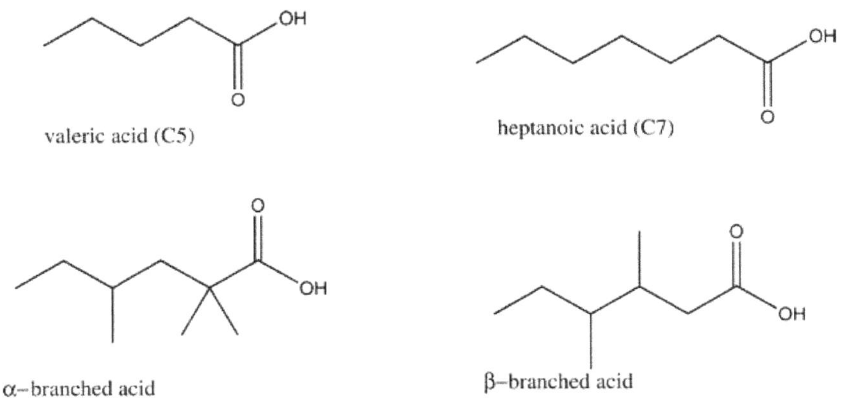

Figure 2. Some of the acids used in the preparation of synthetic lubricants.

2.2. Common Additives

Lubricants with ester basestocks require a series of additives in order to lubricate under the conditions observed in turbine engines. Typical additive packages include antioxidants, typically an aromatic amine, an anti-wear additive, typically a phosphate ester and possibly an antifoaming additive and a viscosity index modifier. The structures of various additives are shown in Figure 3.

Most additives degrade as a part of their mechanism of action, which means that their concentration is constantly decreasing. Many of them also degrade though other mechanisms as well. In general, when the additives have degraded beyond a certain point, either they must be replenished or the lubricant must be changes.

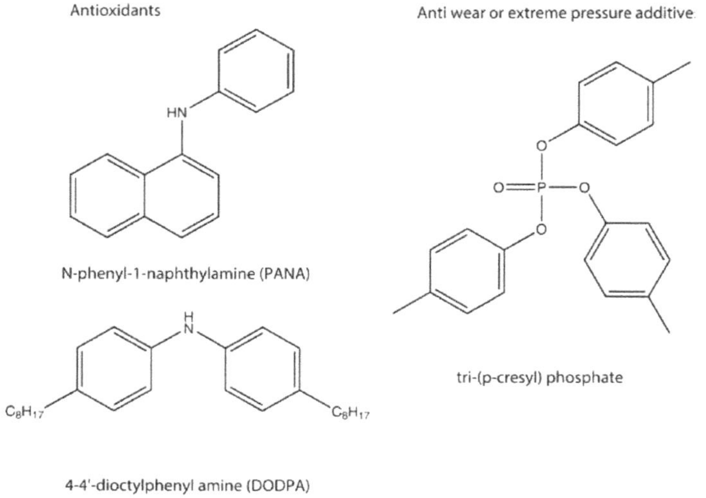

Figure 3. Structures of some lubricant additives used for turbine engines.

Fortunately, most turbine engines lose some lubricant under normal operating conditions and the oil lost is replenished on a regular basis. These procedures maintain the additive packages at acceptable levels.

3. LUBRICANT BASESTOCK DEGRADATION MECHANISMS

Conventional lubricants are petroleum based and consist of hydrocarbons including a huge number of isomers. The primary degradation mechanism for hydrocarbons is oxidation, which leads to the formation of alcohols and carboxylic acids. Synthetic lubricants typically by oxidation to carboxylic acids, aldehydes and ketones under extreme conditions, and degrade by hydrolysis, due to the presence of water and in some cases by transesterification with phosphate ester additives. In addition to the degradation of the basestock due to oxygen, the role of bearing surfaces where extremely high temperatures and pressures; along with the presence of metals and surface treatments such as metal carbides most be considered. In addition, lubricant esters can act synergistically with certain additives [6] and can react differently in the present of metals and or metal carbides.

3.1. Hydrolysis

The hydrolysis of the ester basestock is the reaction of the basestock with water to form an alcohol and a carboxylic acid. This reaction is catalyzed by acids or bases, which are frequently present within the lubricant and does require water. The water can come from various sources, including contamination of the lubricant and the exposure of the lubricant to the environment. Water is soluble in typical ester basestocks to a level of about 500 ppm, meaning that water is readily available in the lubrication systems for turbine engines. The mechanism for the hydrolysis of esters is shown in Figure 3.

Hydrolysis of esters can occur through either an acid or base catalyzed mechanism, with significant differences in the mechanism. The acid catalyzed mechanism [7] begins with the protonation of the carbonyl oxygen atom, followed by a water molecule attacking the carbonyl carbon atom of the ester. The carbonyl carbon normally has a partial positive charge which is increased by the protonation of the oxygen atom yielding the hemiacetal shown in Figure 4. One of the water can be transferred to the alcohol oxygen atom and then the alcohol is lost completing the hydrolysis.

The base catalyzed mechanism [8] involves a water molecule attacking the carbonyl carbon atom, followed by transfer of a proton to the carbonyl oxygen atom. The base the assist with the transfer of the proton from the carbonyl oxygen atom the oxygen atom of the alcohol as the alcohol leaves forming the carboxylic acid.

The two hydrolysis mechanisms require that water be able to attack the carbonyl group of the ester. The use of hindered alcohols such as the various neopentyl alcohols (Figure 1) reduces the ability of the water to approach the carbonyl carbon atom. The use of branched chain acids further reduces the ability of water to attack the carbonyl, resulting in an increase in the hydrolytic stability of the ester [9].

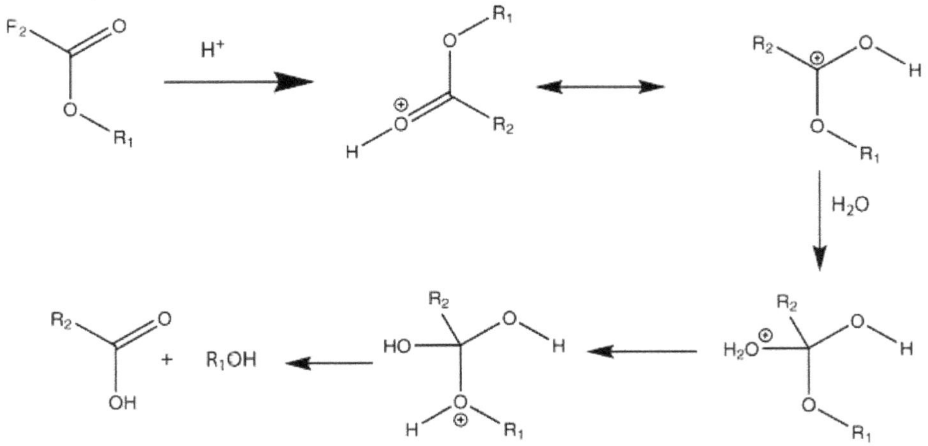

Figure 4. Mechanism for the acid catalyzed hydrolysis of esters.

3.2. Oxidation

Ester based lubricants are all subject to high temperature oxidation which has the most detrimental effect on their properties. Early work examined changes in the bulk composition of ester based lubricants showing the formation of a wide range of acids. The lighter carboxylic acids were attributed to oxidation of the acid chains. Other products were attributed to oxidation of the alcohol [10]. Later work proposed an explanation for oxidation that is based on a radical chain mechanism.

Oxidation occurs through a complex radical chain mechanism which is common to a wide range of organic materials. The initial stages of the oxidation involve the formation of an alkyl peroxy radical by reaction with oxygen. The reaction is propagated by the attack of an alkyl peroxy radical on a methylene group of the ester. The position of the acid, has been shown to be significantly more reactive than other methylene groups in the carboxylic acid [11]. This reaction is significantly hindered in the polyol esters, especially when branched chain acids with branches at C-2 are included. A more recent study, using isotope labelling techniques has shown that the initial site of oxidation is at C-1 of the alcohol, cleaving the carbon–oxygen bond between the first carbon of the alcohol and the ester oxygen, followed by further oxidation at that carbon to form the organic acid [12]. After the initial attack, the reaction can progress to form anhydrides which continue to react to form aldehydes, acids and eventually high molecular weight compounds which can form sludge in the engine. The mechanism of the initial stages of the oxidation of the esters is shown in Figure 5.

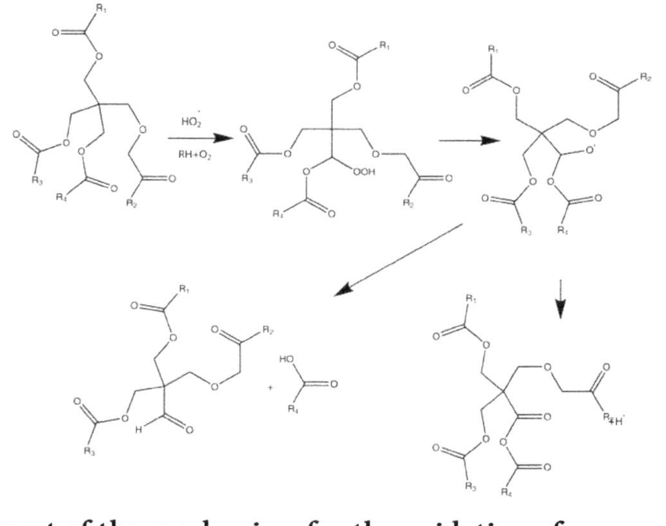

Figure 5. A part of the mechanism for the oxidation of neopentyl polyols.

3.3. Elimination Reactions

Ester based lubricants have been observed to decompose One possible reaction of esters is an elimination reaction in which an alkene and a carboxylic acid are the products. The mechanism for this reaction involves the loss of a proton on the carbon atom leading to the formation of a double bond and the elimination of the carboxylate anion. The mechanism for the elimination reaction is shown in Figure 6.

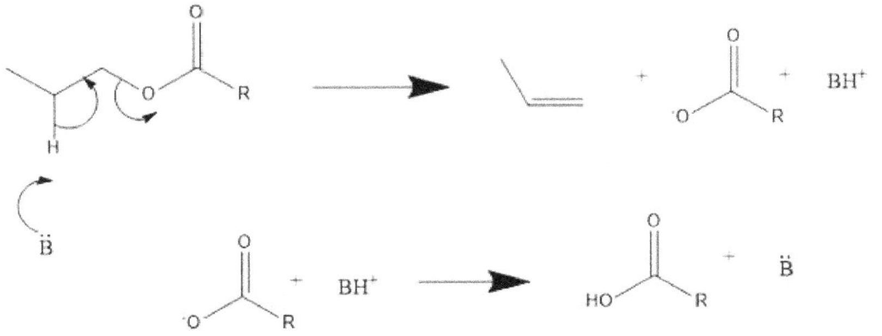

Figure 6. Mechanism for the β elimination reaction.

The use of alcohols without hydrogen atoms at the carbon atom eliminates this mechanism, but under operating conditions of turbine engines, high temperature and metal catalyzed elimination reactions are possible. For this reason, modern ester based lubricants are based on neopentyl polyols, where elimination is blocked due to the lack of hydrogen atoms at the position. Significant work has been conducted on optimizing the properties of the lubricant for use in turbine engines [13, 14].

3.4. Role of Bearing Materials as Catalysts

Lubricant basestocks, in addition to being subjected to high temperatures and pressures, are also in contact with bearing surfaces which contain a combination of metals, metal oxides and surface carbides. Under normal circumstances, ferrous metal are known to increase the rate of thermal degradation of polyol ester based lubricants, especially at temperatures above 220°C. The mechanism for this reaction, however is not completely understood [15]. The incorporation of phosphate esters is known to reduce the catalytic effect of ferrous metals, probably due to the formation of a phosphate film on the surface of the metal (see Section 4.1.2) [16].

4. LUBRICANT ADDITIVE DEGRADATION

Lubricant additives are in many ways designed to degrade as they serve their purpose in the formulated lubricant. As the lubricant is lost in service primarily due to leakage, new lubricant is added which act to replenish the additives used. Lubricant loss is typically estimated at as much as one quart per hour depending on the engine [17]. It is possible to use the amount of remaining additives to determine the need for engine service or lubricant replacement. One example of an instrument for the analysis of remaining antioxidant as an engine diagnostic is RULER [18].

4.1. Phosphate Esters

Phosphate esters are normally required as an extreme pressure or anti-wear additive. The phosphate esters react with the metal surface to form a lubricious polymeric coating. The coating protects the bearing under conditions of start-up, inadequate flow or extreme shear, where the coating wears away, but is continuously reformed from unreacted phosphate ester in the lubricant. The mechanism of action of the additive causes its degradation over time [19].

4.1.1. Hydrolysis

Hydrolysis of phosphate esters is the reaction of the triester with water to form a diester and an aromatic alcohol. The diester can further react under the same conditions to form the monoester and eventually phosphoric acid. Two classes of mechanisms have been proposed for the hydrolysis in aqueous solution, dissociative mechanisms that proceed through a PO3- anion and associative mechanisms then proceed through a penta coordinate phosphorous intermediate [20]. The likely mechanism in the non-polar lubricant medium where the attacking species is a water molecule is most likely through the associative mechanism, which does not require formation of a PO3- ion. The mechanism of the reaction is based on the addition of water to the phosphorus atom, followed by loss of a proton and elimination of the alcohol (phenol) [21]. The mechanism is shown schematically in Figure 7.

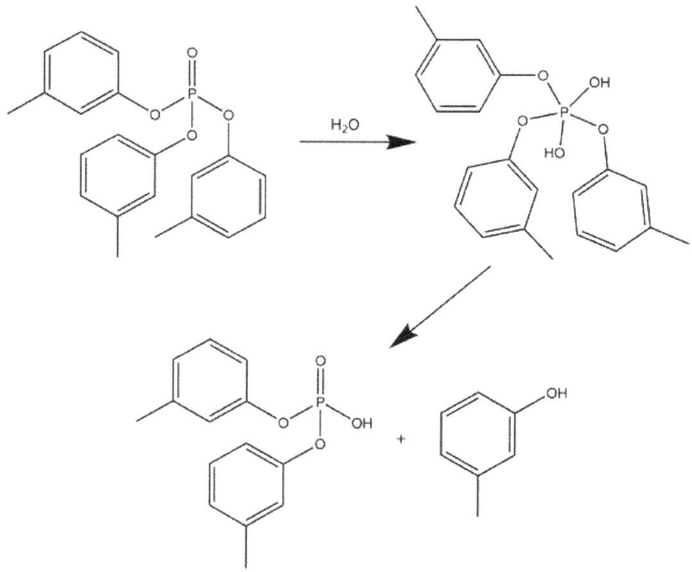

Figure 7. Mechanism for the hydrolysis of phosphate esters in polyol ester-based lubricants.

Hydrolysis is an important degradation mechanism because it forms a range of phosphate partial esters, some of which do not form a lubricous coating on the bearing and contribute to the acids contained in the lubricant.

4.1.2. Polymerization at Metal Surface

Phosphate esters are used as anti-wear of extreme pressure lubricants and work by reactions with the bearing surface to form a polymeric coating that is durable and lubricious [22]. The reaction normally occurs at the oxidized metal surface and results in the formation of an initial layer of graphite, followed by a layer of an iron rich, iron polyphosphate [23]. After the initial coating is formed the film can increase in thickness as iron diffuses to the surface [24]. The coating continuously wears away during use and is reformed as iron diffuses through the coating. The nature of the polymeric lubricous film is shown in Figure 8.

The mechanism for the formation of a polyphosphate polymer begins with the bonding of the phosphate ester (typically tricresylphosphate) to the oxidized iron surface, displacing cresol. The initial steps of the mechanism that leads to the formation of a coating is shown in Figure 9.

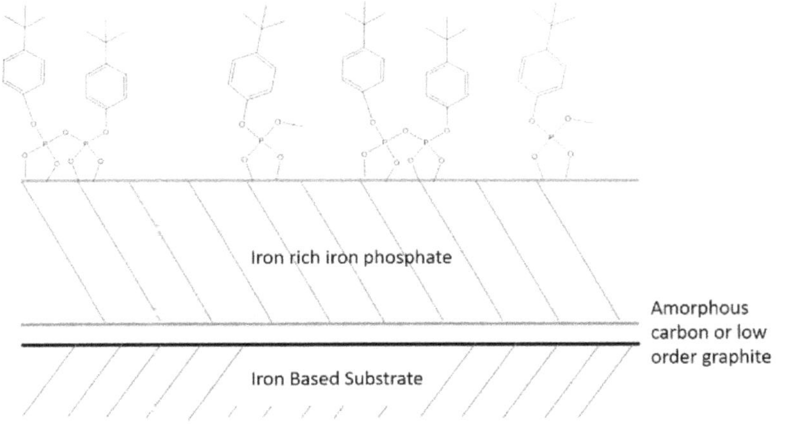

Figure 8. Schematic representation of the iron phosphate film.

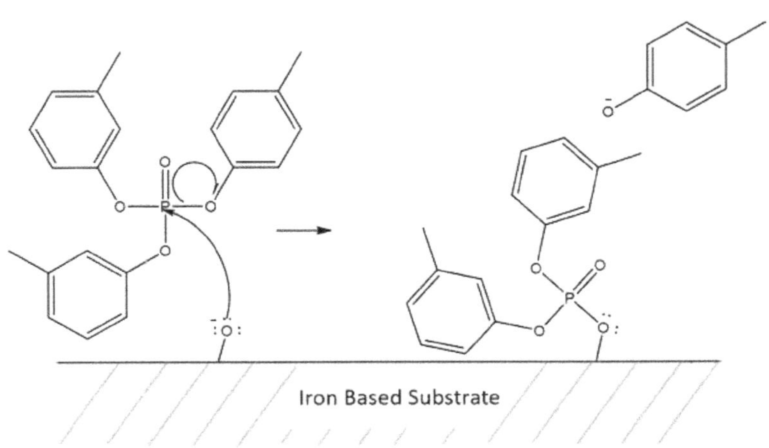

Figure 9. Mechanism of phosphate film formation and structure of phosphate film.

The bound phosphate reacts further with other bound phosphate esters displacing additional cresol leading to the formation of a polymeric coating strongly bound to the metal surface. Typically, on the surface of the metal some of the partially reacted phosphate remains. X-ray photoelectron spectroscopy results show a surface composition corresponding to approximately one cresol remaining per phosphorus atom on the surface as is shown in Figure 8.

Under extreme pressure conditions, the outer layers are removed from the surface and are lost as polymeric phosphorus containing nanoparticles which are not reconverted to the triaryl phosphate in the lubricant. It should be noted that this mechanism explains how the phosphate esters act as an anti-wear additive but it also leads to the degradation of the phosphate esters. The formation of nanoparticles through the wear of the coating formed at the bearing surface leads to a darkening of the oil color, but many of these particles are remove by filtration or eventually settle in the oil sump.

4.1.3. Trans Esterification Lubricant Esters

The last of the reactions of phosphate esters is the reaction between phosphate esters and lubricant esters to form aryl esters and alkyl phosphate esters. This is a reaction that can occur in either a single step or could initially for the acid which can further react to form another ester. The single step process is shown in Figure 10.

This reaction can be of particular concern since the alkyl phosphate formed can undergo transesterification intra molecularly to form the product shown in Figure 11 which is structurally similar to the known neurotoxin which would be formed by a similar reaction with trimethylolpropane [25, 26].

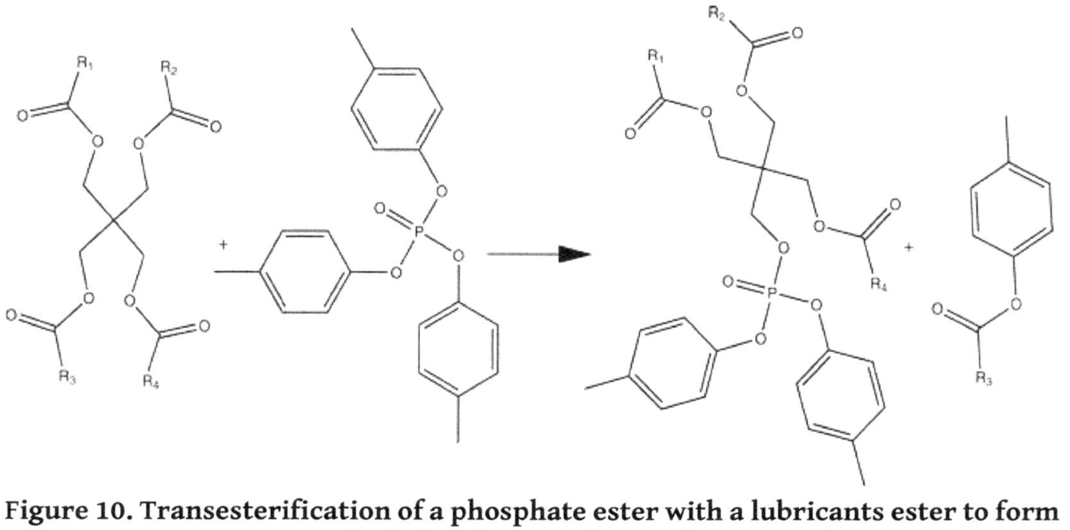

Figure 10. Transesterification of a phosphate ester with a lubricants ester to form an alkyl phosphate and an aryl ester.

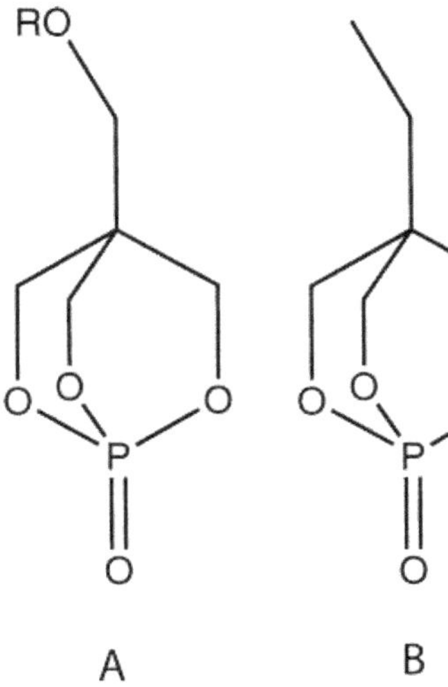

Figure 11. Final product of the transesterification of pentaerythritol ester (A) and the known neurotoxin formed from trimethylol propane (B).

The structure shown in Figure 11(A) assumes the final acid group has been hydrolyzed. Either this compound or the corresponding ester might be assumed to have a toxicity comparable or greater than the compound shown in Figure 11(B).

4.1.4. Addition to Pendant Groups

A final reaction that occurs with phosphate ester additives is addition reactions on the pendant aromatic rings. In this reaction, the carbon-oxygen bond in a phosphate ester is broken at the metal surface. The leaving group remains at the metal surface until it is added to another molecule of phosphate ester [27]. The mechanism for the formation of addition products is shown in Figure 12.

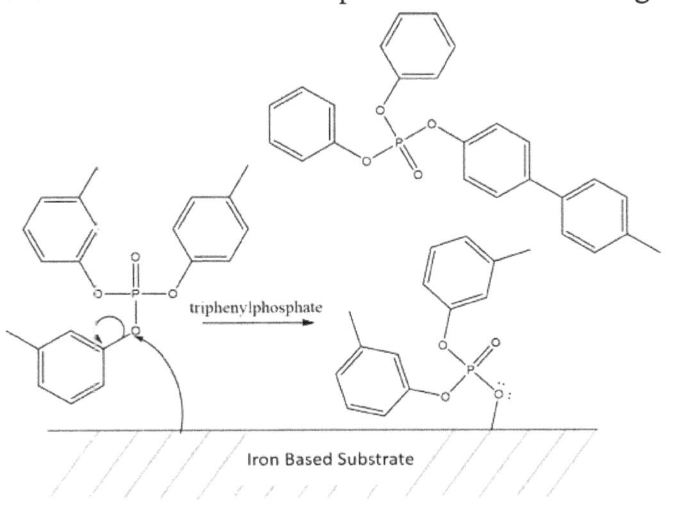

Figure 12. Reaction of phosphate esters with reduced metal surfaces showing the addition of a tolyl group to triphenyl phosphate.

These addition reaction result in higher molecular weight species that might in part be responsible for the formation of the layer of carbon, initially described as a carbide layer [28], but later determined to be either amorphous carbon or low order graphite [29], immediately adjacent to the iron surface. This layer is consistently observed in Auger spectroscopy as is shown in Figure 13.

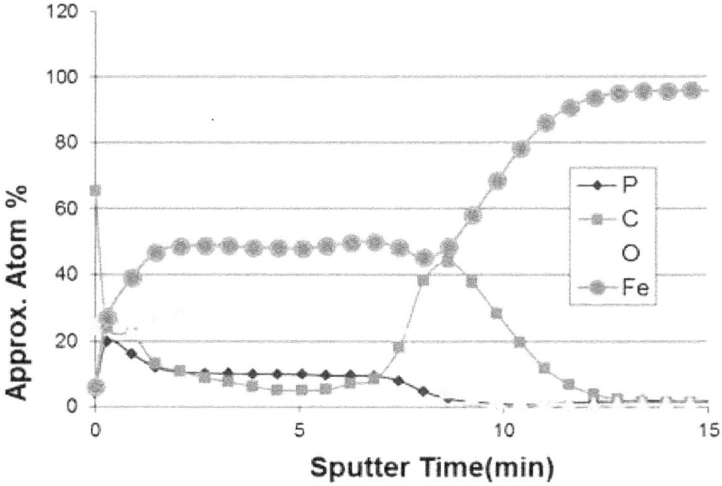

Figure 13.Auger depth profile of a film formed by the deposition of BTPP onto an iron foil at 425°C under nitrogen (sputter rate 1.5 nm/min).

4.2. Antioxidants

Synthetic lubricants are oxidative degraded via a radical chain mechanism at high temperatures. Molecular oxygen abstracts a hydrogen atom forming a free radical. The radical reacts with the basestock abstracting hydrogen atoms or other groups, adding that fragment and creating a new radical and in general increasing the size of the molecule. The chain mechanism continues until the growing chain encounters another radical, resulting in chain termination. Antioxidants are typically added to the lubricant formulation to reduce the rate of lubricant decomposition by reacting with radicals formed in the initiation step of lubricant oxidation.

Anti-oxidant additives can act in two different ways. First, they can react with oxygen to form a stable species reducing the possibility of the chain initiation step in the mechanism. Second, the antioxidant can react with radicals formed, forming a more stable species and acting as a chain termination step [30]. Among the most common types of antioxidants used in lubricants are hindered phenols and aromatic amines.

Aerospace lubricants typically rely on the hindered aryl amines N-phenyl-1-naphthylamine (PANA) and p-dioctyldiphenyl amine (DODPA) (structures shown in Figure 3) as antioxidants because they have the potential to react with a greater number of hydroperoxy radicals [31]. There are two very common mechanisms in which aryl amines act as antioxidants, a low temperature (<120°C) and a high temperature mechanism (>120°C). A common feature of the mechanisms is the reaction of the amine to form radicals. These reactions form aminoxy radicals to form N-alkoxyamines which appear to be the actual antioxidant species [32]. The high temperature mechanism through which aryl amines act as antioxidants is shown in Figure 14.

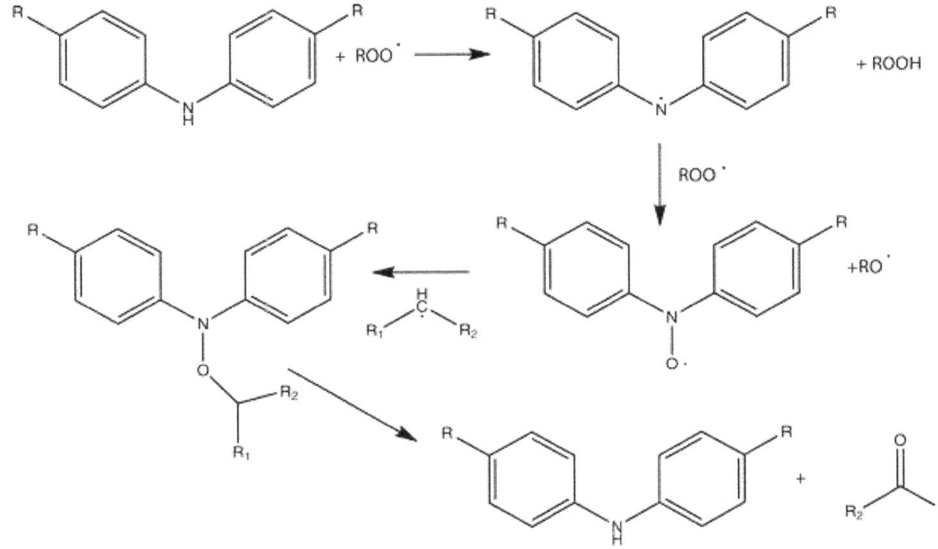

Figure 14. High temperature mechanism for the antioxidant activity of alkylated diphenyl amine antioxidants.

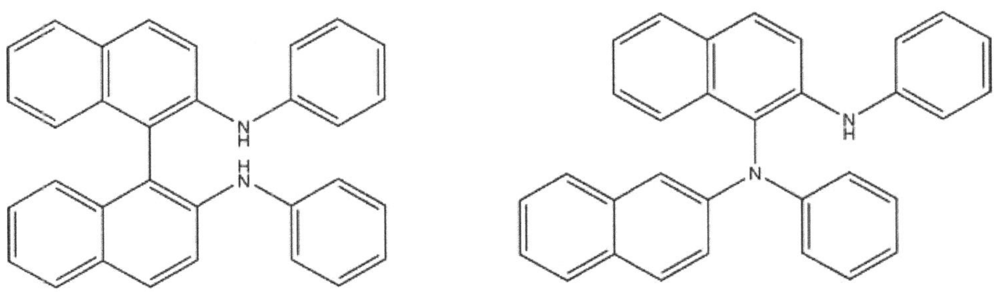

Figure 15. Products of the reaction of PANA as an antioxidant in lubricants.

Other mechanisms that have been reported examined the possibility that the diphenyl amine radical formed in the first step in Figure 14 could disproportionate and then react with itself to form more complex species that eventually lead to poly conjugated systems upon reaction with additional hydroperoxy radicals. The reaction of N-phenyl-1-naphthylamine proceeds somewhat differently due to the susceptibility of the hydrogen of the naphthyl ring to radical attack leading to the formation of dimers and higher polymers as in Figure 15 [33] or the formation of quinone imines and naphthoquinones [34].

5. SYNERGISTIC REACTIONS BETWEEN LUBRICANTS, ADDITIVES AND BEARING MATERIALS

The reactions of the individual components are not always sufficient to predict the chemistry of a formulated lubricant. Some reactions are inhibited by the additives, but may be accelerated by combinations of additives and surface chemistries. Rolling contact fatigue testing with M-50 bearings, for example indicated that PANA and DODPA added to a lubricant along with tricresyl phosphate resulted in an increase in wear over systems where the PANA and DODPA were absent [35]. An explanation might include the antioxidants reduce the oxidation of the metal surface which interferes with the binding of the phosphate ester to the surface [36]. Another system where results are unpredictable is when advanced bearing materials are used with polyolesters and phosphate esters. These observations demonstrate the importance of considering all of the components in the lubrication system instead of the individual reactivities of the various components.

6. INCORPORATION OF ADVANCED BEARING STEELS, IONIC LIQUID ADDITIVES AND NANOPATRITLE BASED ADDITIVES

6.1. Advanced Bearing Steels and Ceramic Bearings

The need for more efficient and more powerful jet engines for military and commercial applications has caused a need for lighter, more durable bearing materials. Harder metal alloys and ceramic bearings are approaches to serve these needs. Changes in bearing materials, however may not be completely compatible with current lubricant basestocks and additive packages.

Many advanced bearing materials are made from carburized stainless steels. The materials begin with a stainless steel which can be formed into the desired shape. The part is then heat treated in the presence of a carbon source resulting in the formation of surface carbides [37]. The surface carbides increase the hardness of the surface significantly. Phosphate esters have been shown to interact with the stainless steels in the absence of carburization [38], but in the presence of all three components, metal carbides, phosphate esters and polyol esters the decomposition is much more rapid [39]. When carburized bearings were tested with polyol ester based lubricants formulated with phosphate esters, an increase in fatigue life and wear performance was observed [40].

Ceramic bearings have good potential for high temperature use in turbine engines. Unlubricated ceramic bearings performed poorly, however when an appropriate lubricant was added they performed better [41]. Typical lubricant additives, however did not perform well under conditions typically seen in steel bearings. At very high temperatures, a film was formed but it did not decrease friction or increase bearing life [42]. To form a lubricious anti-wear coating, the ceramics were pretreated to introduce a thin film of iron which allowed the phosphates to from an anti-wear coating [43].

6.2. Ionic Liquid Additives

Ionic liquids have been considered as potential replacements for both basestocks and additives. As a potential replacement for the basestock, increased costs make them inappropriate for use in turbine engines [44]. A number of ionic liquids are under investigation for use as anti-wear of extreme pressure additives. These additives contain phosphorus in either the cation, as a phosphonium ion or the anion as a tri-alkyl phosphate. Ionic liquids that incorporate the phosphorus in the phosphate anion have been shown to be the most effective [45]. Ionic liquids containing tri-alkyl phosphates interact strongly with metal surfaces through mechanisms also seen in the tri-aryl phosphates [46] discussed in Section 3.1. Ionic liquids with phosphonium cations with a non-phosphate anion have shown superior performance under high load [47]. Ionic liquids have the advantage of reduced volatility, which is important in some applications. Ionic liquid based anti-wear additives show some of the same interferences with antioxidants that are observed with triaryl phosphates, performing better in oils where the additives have been depleted [48].

6.3. Nanoparticle Based Lubricant Additives

Nanomaterials and nanoparticles have been studied for use as additives in liquid lubricants. Some of the initial problems that have been discovered are the dispersion of the nanoparticles and the stability of the dispersion. Capping metal nanoparticles with a monolayer of non-polar organic molecules have resulted in nanoparticles that are oil soluble [49]. A wide range of nanoparticles have been studied and several have shown promise for use in liquid lubricants. Chemical composition was found to be important in anti-wear performance, where morphology and size of the particles were more important in friction reduction. Nanoparticles with layered structures were among the better morphologies [50]. Nanomaterials as lubricant additives appear to have a bright future in lubrication, although none are in current use in aerospace liquid lubricants.

7. CONSEQUENCES OF LUBRICANT DEGRADATION

Lubricant degradation has a significant effect on the properties of the lubricant which can have significant consequences in aerospace. Degradation results in an increase in the chemical reactivity of the oil through the formation of acid and bases, changes in viscosity and changes in thermal conductivity. All of these can result in reduced life of the engine and also decreased operational efficiency. It is important that all of these effects be minimized for safe air travel.

There is an additional safety concern associated with lubricants and their degradation products present in most commercial and military aircraft. Air used to pressurize the cabin is drawn from the engine through a bleed air nozzle. While under normal operation, the air is thought to be safe, seal leakage results in traces of lubricant directed into the cabin. In cases of seal failure, high concentrations of lubricants, additive and degradation products enter the cabin. Smoke events are caused by seal failures, as well as other causes. Fume events occur in 2.1 of every 10,000 flights [51] and oil fumes are noted in 1% of all flights. The health related concerns are indicated by the 30% of fume events where crew impairment has been recorded even though there is recognized under reporting of impairment [52].

Aerotoxic syndrome has been described as an occupational illness along with epidemiological evidence [53]. Possible toxicological mechanism leading to aerotoxic syndrome has been described by Howard et al. [54]. A possible cause for Aerotoxic syndrome is based on repeated low dose exposure to organo-phosphorus compounds derived from phosphate esters [55]. High doses of organophosphates are known to cause organophosphate induced peripheral neuropathy (OPIDN) [56], however the doses encountered here are much lower, suggesting other chronic mechanisms [57]. The toxicity evidence indicates the need for clean air requirements for aircraft using bleed air for cabin pressurization [58].

8. CONCISE SUMMARY

The sections above have covered the typical processes that cause deterioration in lubricant basestocks and additives. The deterioration of the basestock is the primary worry for the overall health of the engine. The breakdown of the additives has a significant role in their functionality. Significant work has been dedicated to identifying additives that react suitably, have low volatility, and are thermally stable. The additives in the lubricant degrade with time but will continue to operate as long as they are not totally used up. By designing the molecular structure of the esters in the basestock, the significance of certain mechanisms has been diminished. Contemporary esters in the base oil are derived from polyols that lack hydrogen atoms in the position, preventing removal by this process. Hydrolysis is a major concern because it generates acids and changes the physical characteristics, such as viscosity and pour point, of the lubricant. Oxidation can generate acids and alter the lubricant's physical qualities. Improved antioxidants have diminished the significance of this pathway. The acids created by oxidation or hydrolysis are carboxylic acids, which are less corrosive than the mineral acids resulting from the oxidation of sulfur and nitrogen components in mineral oils. New lubricants are being developed to raise the working temperature without experiencing a notable decline in their qualities. Molecular design has been utilized to inhibit basestock degradation mechanisms by selecting certain acids to create polyol esters that can prevent oxidation and hydrolysis. Understanding these mechanisms has enabled the development of high-performance lubricants.

9. AT A GLANCE

Ester-based synthetic lubricants have been utilized in different formulations along with antioxidants, phosphorus-based anti-wear additives, and other additives for an extended period. Both the basestock and additives undergo changes in their physical and chemical properties with use. Basestocks typically deteriorate through several ways, whereas additives can decay or be consumed as they react to fulfill their intended purpose. This chapter will analyze the formulation of contemporary turbine engine lubricants and the processes via which these lubricants deteriorate with time. Future engines are considering new bearing materials and the impact of potential ionic liquids additives on current additives. The presentation will cover the influence of degradation on lubricant characteristics, its impacts on turbine engines, and its potential implications for human health. These novel materials have many potential degradation pathways that need to be assessed prior to their widespread adoption.

REFERENCES

1. Johnson DW. Lubricants for turbine engines. In: Agarwal RK, editor. Recent Progress in Some Aircraft Technologies. Rijeka, Croatia: InTech; 2016. pp. 35-54. DOI: 10.5772/62394
2. Howard CV, Johnson DW, Morton J, Michaelis S, Supplee D, et al. Is a cumulative exposure to a background aerosol of nanoparticles part of the causal mechanism of aerotoxic syndrome? Journal of Nanomedicine and Nanoscience. 2018;2018:139. DOI: 10.29011/JNAN-139. 100039
3. Lansdown AR. Aviation lubricants. In: Mortimer RM, Orszuliz ST, editors. Chemistry and Technology of Lubricants. Boston: Springer; 1994. pp. 223-236. DOI: 10.1007/978-1-4615-3554-6_9
4. MIL-PRF-23699G, performance specification: Lubricating oil, aircraft turbine engine, synthetic base, NATO code numbers: O-152, O-154, O-156, and O-167 (13-MAR-2014) 23699 specs. 2014. Available from: http://everyspec.com/MIL-PRF/MIL-PRF-010000-29999/MIL-PRF-23699G_50301/ [Accessed: Jan 31, 2016]
5. MIL-PRF-7808L, performance specification: Lubricating oil, aircraft turbine engine, synthetic base (02 May 1997) [Superseeding MIL-L-7808K]. 1997. Available from: http://everyspec.com/MIL-PRF/MIL-PRF-000100-09999/MIL-PRF-7808L_5699/ [Accessed: Jan 31, 2016]

6. Chao TS, Hutchison DA, Kjonaas M. Some synergistic antioxidants for synthetic lubricants. Industrial and Engineering Chemistry: Product Research and Development. 1984;23:21-27. DOI: 10.1021/i300013a005
7. Roberts I, Urey HC. The mechanisms of acid catalyzed ester hydrolysis, esterification and oxygen exchange of carboxylic acids. Journal of the American Chemical Society. 1939;61(10):2584-2587. DOI: 10.1021/ja01265a003
8. Jencks WP, Carriuolo J. General Base Catalysis of Ester Hydrolysis. Journal of the American Chemical Society. 1961;83(7):1743-1750. DOI: 10.1021/ja01468a044
9. Beimesch BJ, Schnur NE, Hughes CJ Jr. Di- and tripentaerythritol esters of isostearic acid. US Patent 4477383A. Asssignee: National Distillers and Chemical Corp, Cognis Corp; 1984
10. Zeeman A, Koch K, Bartl P. Thermal oxidative ageing of neopentylpolyol ester oils: Evaluation of thermal-oxidative stability by quantitative determination of volatile ageing products. Journal of Synthetic Lubricants. 1985;1:2-21. DOI: 10.1002/jsl.3000020102
11. Bakunin VN. A mechanism of thermooxidative degradation of polyol ester lubricants. Journal of Synthetic Lubrication. 1992;9:127-143. DOI: 10.1002/jsl.3000090204
12. Frauscher M, Besser C, Allmaier G, Dorr N. Oxidation products of ester-based oils with and without antioxidants identified by stable isotope labelling and mass spectrometry. Applied Sciences. 2017;7:397-414
13. Niedzielski EL. Neopentyl polyol ester lubricants-bulk property optimization. Industrial & Engineering Chemistry Product Research and Development. 1976;15:53-58. DOI: 10.1021/i360057a010
14. Beane GA, Gschwender LJ, Snyder CE, Shimski JT. Military aircraft propulsion lubricants–current and future trends. Journal of Synthetic Lubrication. 1986;3:263-291. DOI: 10.1002/jsl.3000030403
15. Kauffman RE. Mechanism for the reaction between polyolester lubricant and ferrous metals, Part 1: Literature search. ASRAE Transactions: Research. 2005;111:378-386
16. Jones RL, Ravner H, Cottington RL. Inhibition of iron catalyzed neopentyl polyol ester thermal degradation through passivation of the active metal surface by tricresyl phosphate. ASLE Transactions. 1970;13:1-10. DOI: 10.1080/05698197008972276
17. Exxon. Jet Oil Consumption Factors. Available from: https://www.exxonmobil.com/en/aviation/knowledge-library/resources/jet-engine-oil-consumption [Viewed: Oct 28, 2018; Posted: Nov 15, 2016]

18. Myle B, Kauffman R. The use of cyclic voltammetric antioxidant analysis for proactive engine oil condition monitoring program. SAE Technical Paper 2000-01-1810. 2000. DOI: 10.4271/2000-01-1810

19. Johnson DW, Hils JE. Phosphate esters, thiophosphate esters and metal thiophosphates as lubricant additives. Lubricants. 2013;1:132-148. DOI: 10.3390/lubricants1040132

20. Florian J, Wardhel A. Phosphate ester hydrolysis in queous solution: Associative versus dissociative mechanisms. The Journal of Physical Chemistry. 1998;102:719-734. DOI: 10.1021/jp972182y

21. Cox JR Jr, Ramsey OB. Mechanisms of nucleophilic attack in phosphate esters. Chemical Reviews. 1964;64:317-352. DOI: 10.1021/cr60230a001

22. Makki J, Graham EE. Formation of solid films from the vapor phase on high-temperature surfaces. Lubrication Engineering. 1991;47:199-206

23. Saba CS, Forster NH. Reactions of aromatic phosphate esters with metals and their oxides. Tribology Letters. 2002;12:135-146. DOI: 10.1023/A:1014081523491

24. Forster NH. Rolling contact testing of vapor phase lubricants—Part IV: Diffusion mechanisms. Tribology Transactions. 1999;42:10-20. DOI: 10.1080/10402009908982184

25. Rubey WA, Streibich RC, Bush J, Centers PW, Wright RL. Neurotoxin formations from pilot-scale incineration of synthetic ester turbine lubricants with triaryl phosphate additive. Archives of Toxicology. 1996;70:508-509. DOI: 10.1007/s002040050306

26. Centers PW. Potential neurotoxin formation in thermally degraded synthetic ester turbine lubricants. Archives of Toxicology. 1992;66:679-680. DOI: 10.1007/BF01981509

27. Johnson DW, Morrow S, Forster NH, Saba CS. Vapor-phase lubrication: Reaction of phosphate ester vapors with iron and steel. Chemistry of Materials. 2002;14:3767-3775. DOI: 10.1021/cm010921o

28. Graham EE, Klaus EE. Lubrication from the vapor phase at high temperatures. ASLE Transactions. 1986;29:229-234. DOI: 10.1080/05698198608981682

29. Forster NH. Rolling contact testing of vapor phase lubricants—Part III: Surface analysis. Tribology Transactions. 1999;42:1-9. DOI: 10.1080/10402009908982183

30. Pospisil J. The key role of antioxidant transformation products in the stabilization mechanisms—A critical analysis. Polymer Degradation and Stability. 1991;34:85, 90113-109, 90116. DOI: 10.1016/0141-3910(91)

31. Soleimani M, Dehabadi L, Wilson LD, Tabil LG. Antioxidants classification and applications in lubricants. In: Johnson DW, editor. Lubrication: Tribology, Lubricants and Additives. London, UK: IntechOpen; 2018. pp. 23-42. DOI: 10.5772/intechopen.72621

32. Schwetlick K, Habicher WD. Antioxidant action mechanisms of hindered amine stabilisers. Polymer Degradation and Stability. 2002;79:35-40. DOI: 10.1016/S0141-3910(02)00116-7

33. Mousavi P, Wang D, Grant CS, Oxenham W, Hauser PJ. Effects of antioxidants on the thermal degradation of a polyol ester lubricant using GPC. Industrial& Engineering Chemistry Research. 2006;45:15-22. DOI: 10.1021/ie050539b

34. Pospisil J. Aromatic and heterocyclic amines in polymer stabilization. Advances in Polymer Science. 1995;124:87-189. DOI: 10.1007/BFb0025229

35. Trivedi HK, Forster NH, Saba CS. Rolling contact fatigue testing of a 3 cSt polyolester lubricant with and without TCP and DODPA/PANA at 177°C. Tribology Letters. 2004;16:231-237. DOI: 10.1023/B:TRIL.0000009734.35530.d

36. Johnson DW. The tribology and chemistry of phosphorus-containing lubricant additives. In: Darji PH, editor. Advances in Tribology. Rijeka, Croatia: InTech; 2016. pp. 175-196. DOI: 10.5772/62208

37. Tabet N, Allam I, Yin RC. X-ray photoelectron spectroscopy investigation of the carburization of 310 stainless steel. Applied Surface Science. 2003;220:259-272. DOI: 10.1016/S0169-4332(03)00820-1

38. Johnson DW, Bachus M, Hils JE. Interaction between lubricants containing phosphate ester additives and stainless steels. Lubricants. 2013;1:48-60. DOI: 10.3390/lubricants1020048

39. Johnson DW, Hils JE, Forster N. Interaction of polyol esters and phosphate esters with metal carbides. Tribology Letters. 2011;42. DOI: 223-232. DOI:10.1007/s11249-011-97666-x

40. Trivedi HK, Forster NH, Rosado L. Rolling contact fatigue evaluation of advanced bearing steel with and without the oil anti-wear additive tricresyl phosphate. Tribology Letters. 2011;41:597-605. DOI: 10.1007/s11249-010-9738-6

41. Zaretsky EV. Ceramic bearings for use in gas turbine engines. Journal of Materials Engineering and Performance. 1989;11. DOI: 237-253. DOI:10.1007/s11665-013-0726-5

42. Bertrand PA. Reactions of tricresyl phosphate with bearing materials. Tribology Letters. 1997;3:367-377. DOI: 10.1023/A:1019105811014

43. Graham EE, Morales W, Patel N. Method of providing ceramic article with wear resistant coating. US Patent 5738913. Asssignee: Cleveland State University; 1998

44. Somers AE, Howlett PC, MacFarlane DE, Forsyth M. A review of ionic liquid lubricants. Lubricants. 2013;1:3-21. DOI: 10.3390/lubricants1010003

45. Zhou Y, Dyck J, Graham TW, Lou H, Leonard DN. Ionic liquids composed of phosphonium cations and organophospjhate, carboxylate and sulfonate anions as lubricant antiwear additives. Langmuir. 2014;30:13301-13311. DOI: 10.102/la5032366

46. Barnhill WC, Qu J, Luo H, Meyer HM III, Ma C, Chi M, et al. Phosphonium-organophosphate ionic liquids as lubricant additives : effects of cation structure on physiochemical and tribological characteristics. Applied Materials and Interfaces. 2014;6:22585-22593. DOI: 10.1021/am506702u

47. Somers AE, Howlett PC, Sun J, MacFarlane DR, Forsyth M. Transition in wear performance for ionic liquid lubricant under increasing load. Tribology International. 2010;40:279-284. DOI: 10.1007/s11249-010-9695-0

48. Anand M, Hadfield M, Viesca JL, Thomas B, Hernandez-Battez A, Austen S. Ionic liquid as tribological performance improving additive for in-service and used fully formulated diesel engine lubricants. Wear. 2015;334-335:67-74. DOI: 10.1016/j.wear.2015.01.055

49. Templeton AC, Weulfing WP, Murray RW. Monolayer protected cluster molecules. Account of Chemical Research. 2000;33:27-36. DOI: 10.1021/ar9602664

50. Dai W, Kheireddin B, Gao H, Liang H. Roles of nanoparticles in oil lubrication. Tribology International. 2016;102:88-98. DOI: 10.1016/triboint.2016.05.020

51. Shehadi M, Jones B, Hosni M. Characterization of the frequency and nature of bleed air contamination events in commercial aircraft. Indoor Air. 2015;26:478-488. DOI: 10.1111/ina.12211

52. Michaelis S. Contaminate aircraft cabin air. Journal of Biological Physics and Chemistry. 2011;11:132-145. DOI: 10.4024/41111/11-4-abs1.jbpc.11.04

53. Michaelis S, Burdon J, Howard C. Aerotoxic syndrome: A new occupational disease? Public Health Panorama. 2017;3:141-356

54. Howard C, Michaelis S, Watterson A. The aetiology of 'Aerotoxic syndrome'—A toxico-pathological viewpoint. Open Access Journal of Toxicology. 2017;1:1-3. DOI: 10.19080/OAJT.2017.01.555575

55. Terry AV. Functional consequences of repeated organophosphate exposure: Potential non-cholinergic mechanisms. Pharmacology & Therapeutics. 2012;134:355-365. DOI: 10.1016/j.pharmthera.2012.03.001

56. Carletti E, Schopfer LM, Colletier JP, Froment MT, Nachon F, Weik M, et al. Reaction of cresyl salingen phosphate, the organophosphorus agent implicated in aerotoxic syndrome with human cholinesterases: mechanistic studies employing kinetics, mass spectrometry and X-ray structure analysis. Chemical Research in Toxicology. 2011;24:797-808. DOI: 10.1021/tx100447k

57. de Ree H, van den Berg M, Brand T, Mulder GJ, Simons R, Veldhuijzen van Zanten B, et al. Health risk assessment of exposure to tricresyl phosphates (TCPs) in aircraft: A Commentary. Neurotoxicology. 2014;45:209-215. DOI: 10.1016/j.neuro.2014.08.011

58. Michaelis S. Aircraft clean air requirements using bleed air systems. Engineering. 2018;18:142-172. DOI: 10.4236/eng.2018.104011

5 Chapter

COMPOSITE FUSELAGE EVOLUTION: MANUFACTURING VIEWPOINT

1. INTRODUCTION: AN OVERVIEW

Studying history helps us comprehend the development of contemporary methods used in composite fuselage construction. It also offers a glimpse into the potential future of composite fuselage manufacture. Figure 1 displays a graph illustrating the rise in composites usage in military and commercial aircraft throughout time. At first, carbon fiber reinforced composites (CFRP) were mostly used in non-structural parts like fairings and flight control surfaces in commercial and military aircraft. Structural applications for military aircraft started in the 1980s when composite materials accounted for over 20% of the structure's weight. As the industry advanced, understanding of materials and processes improved, leading to increased cost-effectiveness. This progress allowed commercial aircraft makers to integrate the technology into the current wide-body and other new aircraft.

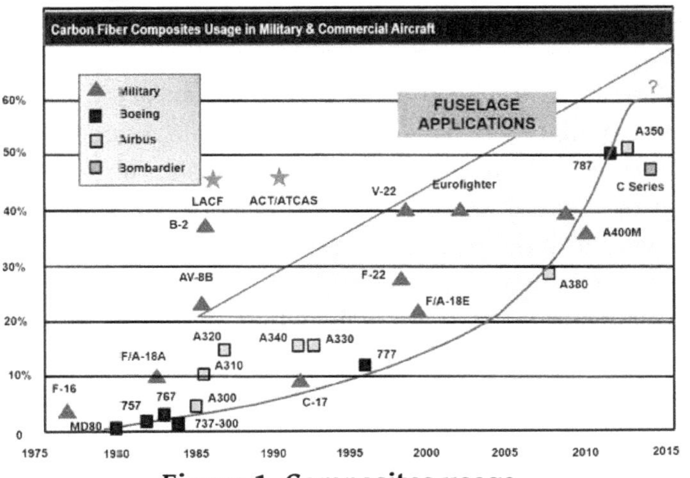

Figure 1. Composites usage.

2. EARLY RESEARCH AND DEVELOPMENT

Research and development of advanced composite materials and methods for aerospace purposes in the United States began in the 1940s at Wright-Patterson Air

Force Base in Dayton, Ohio[1]. The primary focus of this first study was for military purposes. Since then, ongoing research by the Air Force Research Laboratory (AFRL), in collaboration with industry, universities, and government agencies like DARPA and DOE, remains at the forefront of advancing materials for military use. NASA began research in the late 1960s to create advanced composites for use in commercial aircraft and space vehicles. Over time, NASA has collaborated with industry and academics to create cost-effective technologies that enhance the safety and efficiency of airplanes and launch vehicles. The paper NASA Composite Materials Development: Lessons Learned and Future Challenges provides an excellent historical review of NASA's role in the development of composite materials and processes [2].

Both AFRL and NASA supported initiatives used a "building-block" model for research and development, advancing through a succession of steps with increasing complexity and cost, each step building upon the preceding one. Typically, projects began at a coupon level and examined various samples to narrow down design approaches, construction materials, tooling, and manufacturing procedures for building and testing coupons, subcomponents, and eventually full-scale components. This method, similar to the Technology Readiness Levels used for emerging technologies, was effectively utilized in programs like the Air Force's Large Aircraft Composite Fuselage (LACF) Program in the late 1980s and NASA's Advanced Composites Technology (ACT) program in the mid-1990s. The B-2 Stealth Bomber program occurred in the 1980s and offered valuable insights about producing huge composite primary structures. The substantial use of carbon fiber composites in the B-2 was primarily driven by the need for high survivability performance, rather than cost or producibility. Boeing was a key subcontractor in the project and manufactured the wing skins using Automated Tape Laying (ATL). This initiative provided a chance to showcase the efficiency achievable with automated lamination techniques like ATL and AFP. The V-22 program directly benefited from the ACT initiative. Composites are heavily and actively utilized in helicopters more than any other type of aircraft because to the crucial importance of weight. The V-22 incorporates composite materials in many components such as the wings, fuselage skins, empennage, side body fairings, doors, and nacelles. AFP technology is utilized to manufacture the rear fuselage skin as a single unit. Bell and Boeing both use cocured, hat-stiffened fuselage constructions with solid silicone mandrels in their respective parts of the program.

2.1. Large Aircraft Composite Fuselage (LACF) Program

The LACF program was conducted in part by Northrop and was sponsored by the Air Force Wright Aeronautical Laboratory (AFWAL) during the 1980s. The program was part of an effort focused on manufacturing technology for the Linear Manufacturing of Large Aircraft Composite Primary Structure Fuselage. The multi-phase program was directed toward the definition and demonstration of manufacturing methods for cocuring stringer stiffened fuselage panels using (1) existing, qualified material systems; (2) automated skin fabrication; (3) inner mold line (IML) controlled tooling; (4) non-autoclave curing technology. Like many similar terms, in the 1980s "linear" manufacturing was a code word for "lean" and non-autoclave is referred to today as out-of-autoclave or OOA processes.

The program followed a building-block approach through four phases (Figure 2):

- Phase I—methods definition
- Phase II—manufacturing methods establishment
- Phase III—manufacturing verification
- Phase IV—production demonstration

As the program moved through various phases, lessons learned where documented and applied to the next phase. Phase I lessons learned included:

1. Raw material required (tow bad, tape good) changes to improve panel quality using automated lamination equipment
2. Non-autoclave cured panel mechanical properties were equivalent to autoclave cured panels
3. IML tooling is very good at controlling stringer location and dimensions

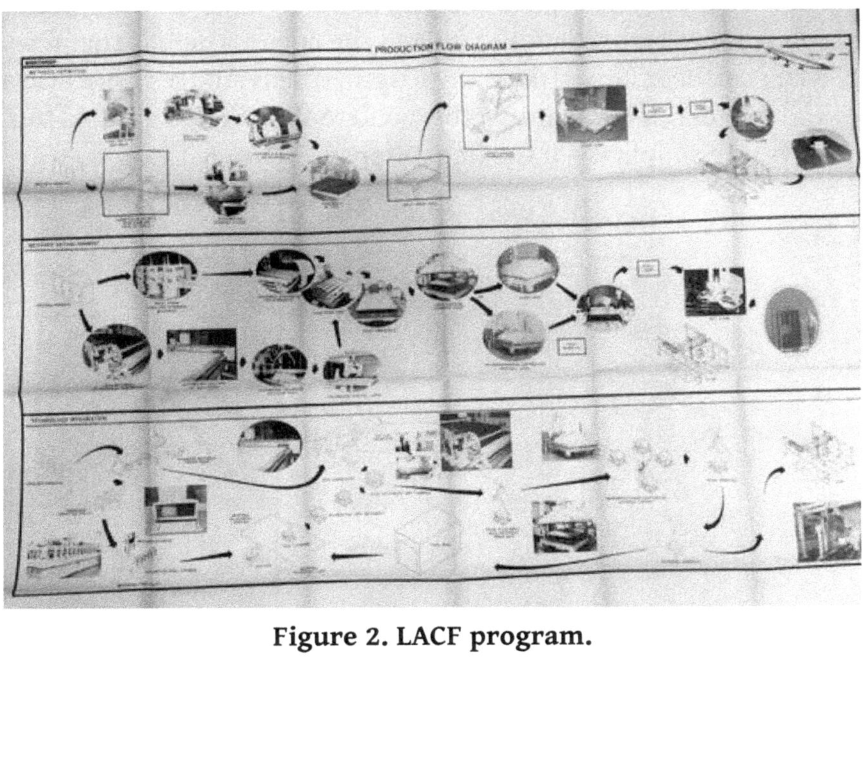

Figure 2. LACF program.

4. IML provides very easy tool loading and bagging
5. Continuous roll forming can be used to preform preplied material into "C" channels ready for tool loading (Figure 3).

Phase II lessons learned included:

1. Non-autoclave cure has risks associated with consumable bagging materials.
2. Integrally heated tooling strongly supports linear manufacturing.
3. Confirmed IML tooling is excellent for controlling stringer/skin dimensions and location.
4. Confirmed IML tooling and "I" beam stringer for part and tool removal.
5. Flat preplied laminates can be drape formed on gentle contours using IML cure tools.
6. Automation can be applied but presents reliability risks and potential equipment downtime.
7. Automation can produce a laminate that does not require additional debulking.
8. Roll forming of stringer "C" channels is important for linear manufacturing (Figure 4).

Among the lessons learned as a result of Phases III and IV were the economics related to process scale up for both size and rate. This included ply cutting and kitting time for panel fabrication and backing paper removal and management issues affecting tow placement and stringer laminate preplying (Figure 5). Another lesson included gaining a better understanding of cocuring longitudinal "I" beams to the skin of a large fuselage panel. One nice feature of the "I" beam construction is that the tooling is not trapped after cure and the channel details that form the "C" of the "I" beam can be removed over any length. Disadvantages were also apparent including the number of laminate preform and tooling details needed to construct an "I" beam vs. the simplicity of the hat stiffener (Figure 6).

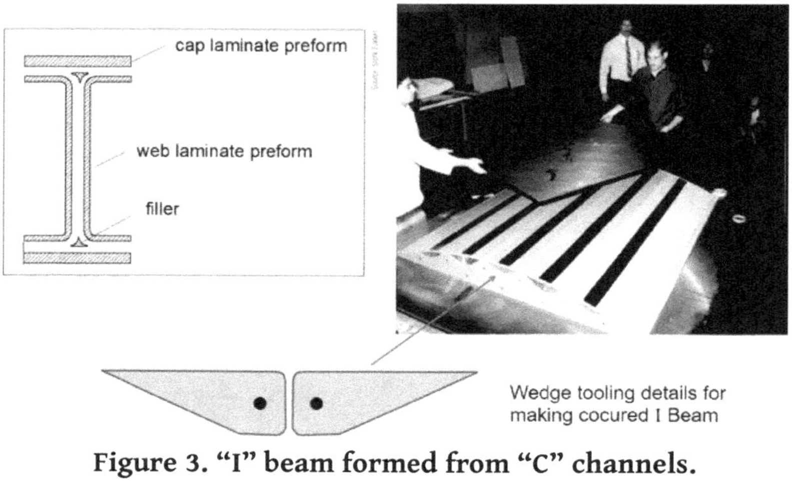

Figure 3. "I" beam formed from "C" channels.

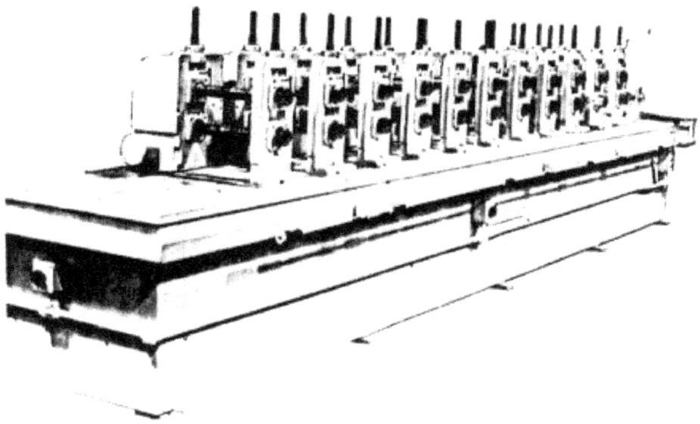

Figure 4. "C" channel roll forming machine.

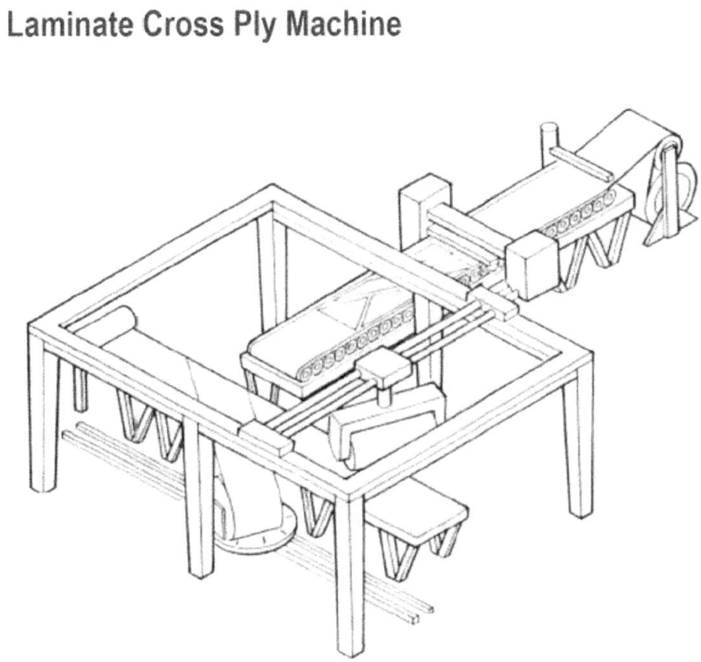

Figure 5. Laminate cross ply equipment.

Northrop developed hat stiffened fuselage skin manufacturing technology in support of the YF-23 (Figure 7). One critical problem to solve was the removal of hat stiffener mandrel tooling from the cured part. The fuselage tooling was OML controlled and constructed from CRFP prepreg to match the coefficient of thermal expansion (CTE) of the parts. The resin system used for the tooling was bismaleimide (BMI) and the tools were autoclave cured on male, machined monolithic graphite source tools. The hat stiffeners that run longitudinally along the skin were cocured using a silicone mandrel system developed by Northrop using Rubbercraft as a supplier.

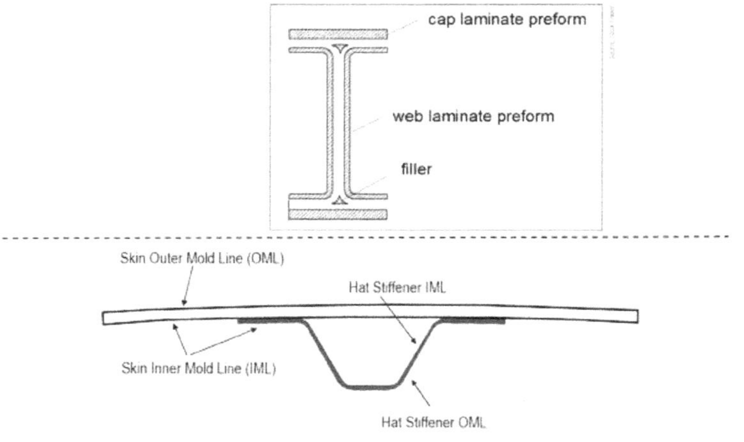

Figure 6. "I" beam vs. hat stiffener.

Figure 7. YF-23 fuselage structure.

The silicone based solid mandrel system included a solid rubber mandrel, a butterfly caul and a resin end dam. The silicone mandrel was designed to be removed from the cured part after pulling and elongating the mandrel to reduce the cross section enough to release from the part. The butterfly caul was designed to help consistently control the OML of the hat stiffener. It also helped to greatly simplify the bagging process which allows for the use of a broader range of operators instead of relying solely on a highly skilled mechanic. The end dam was designed to be cheap and disposable and replace much of the inner bagging process complexity of sealing off the hat stiffener to prevent resin bleed during the cure cycle (Figure 8). This is not a hard process, but is critical and tedious.

Northrop subsequently applied this hat stiffener fabrication process technology to the fuselage of the F/A-18E/F as a prime subcontractor to Boeing on the program (Figure 9).

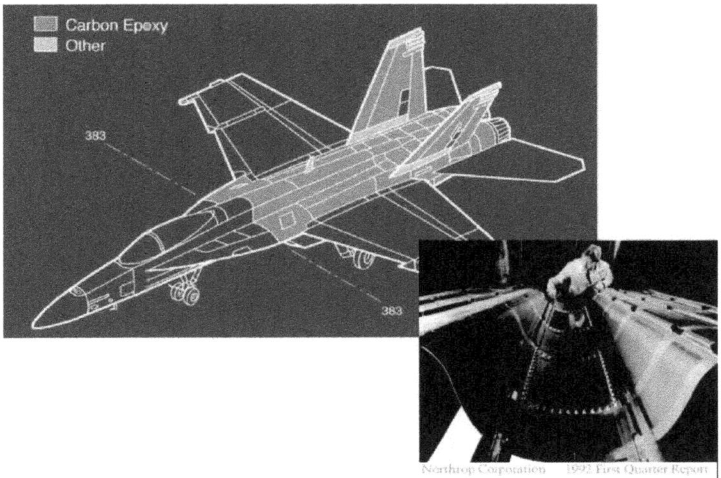

Figure 8. Solid mandrel system.

During this time period, it was recognized by many of the R&D programs that liquid molding processes presented the opportunity to use resins and fibers in their lowest-cost state by eliminating prepreg from the fabrication process. Other advantages included minimizing material scrap, simplifying raw material storage, and supporting non-autoclave fabrication processes.

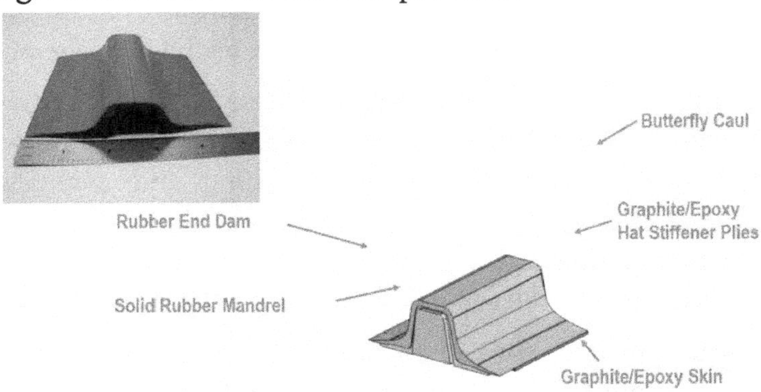

Figure 9. F/A-18E/F fuselage structure.

The development of net shape damage-tolerant textile preforms and the development of innovative liquid molding tooling concepts supported this opportunity. The Advanced Composites Technology (ACT) program included processes such as resin transfer molding (RTM) and pultrusion in the development efforts. The technologies have progressed to state-of-practice processes with both the 787 and the A350 programs using liquid molding and textile preform technology for fabricating fuselage frame elements.

2.2. Advanced Composites Technology (ACT) Program

The objective of the ACT fuselage program was to develop composite primary structure for commercial airplanes with 20–25% less cost and 30–50% less weight than equivalent metallic structure [3]. The Advanced Technology Composite Aircraft Structure (ATCAS) program was performed by Boeing as the prime contractor under the umbrella of NASA's ACT program and focused on fuselage structures. A large team of industry and university partners also supported the program. The primary objective of the ATCAS program was to develop and demonstrate an integrated technology that enables the cost and weight effective use of composite materials in fuselage structures for future aircraft.

The area selected for study was identified as Section 46 on Boeing wide body aircraft (Figure 10). This section contains many of the structural details and manufacturing challenges found throughout the fuselage. This includes variations in design details to address high loads at the forward end and lower portions of the fuselage. The loads decrease toward the aft end and the upper portion of the fuselage, allowing for transitions in the thickness of the structure that are tailored to match the structural loading.

A quadrant panel approach was selected for study as shown in Figure 11. The cross section is split into four segments, a crown, keel, and left and right side panels. The circumferential, four quadrant panel approach was selected with the idea of reducing assembly costs by reducing the number of longitudinal splices. This built-up assembly approach is baseline to metallic aircraft manufacturing and is similar to the approach Airbus selected for most of the fuselage of the A350.

Manufacturing process development and design trade studies contributed to the development of Cost Optimization Software for Transport Aircraft Design Evaluation (COSTADE) which allowed for defining and evaluating the cost-effectiveness and producibility of various designs. Included in the program were assessments of tooling, materials and process controls needed for future full-barrel fabrication like Boeing selected for the 787.

The structural concepts studied included stiffened skin structures achieved by stand alone or combinations of cocuring, cobonding, bonding, and mechanical attachment of stringers and frames to monolithic or sandwich panel skins (Table 1). The crown section study selected fiber placed skins laminated on an IML controlled layup mandrel with the skin subsequently cut into individual panels and transferred to OML cure tools. Hat stiffeners used solid silicone mandrels located longitudinally along the IML of the skin panels for cocuring.

The recommended optimized panel design included cobonding of cured frame elements while cocuring the hat stiffeners and the skin. The cured frames were demonstrated using braided textile preforms and resin transfer molding (RTM).

One of the main challenges of the crown panel concept was the bond integrity between the precured frames cobonded to a skin panel that is stiffened with cocured hat stringers. Alternative concepts the team considered during the review process included mechanically attached Z-section frames instead of cobonded J's.

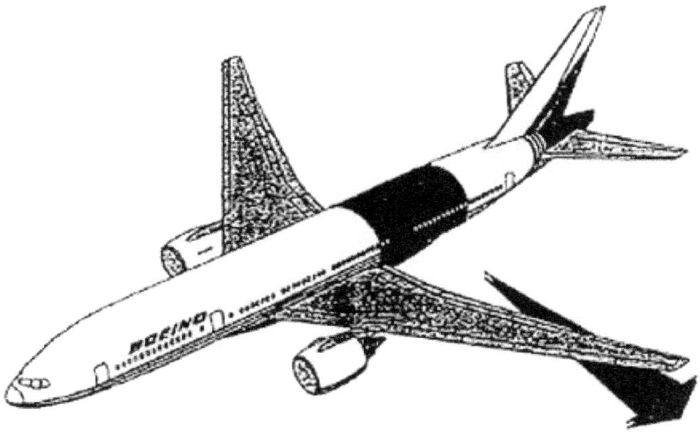

Figure 10. ACT fuselage section [3].

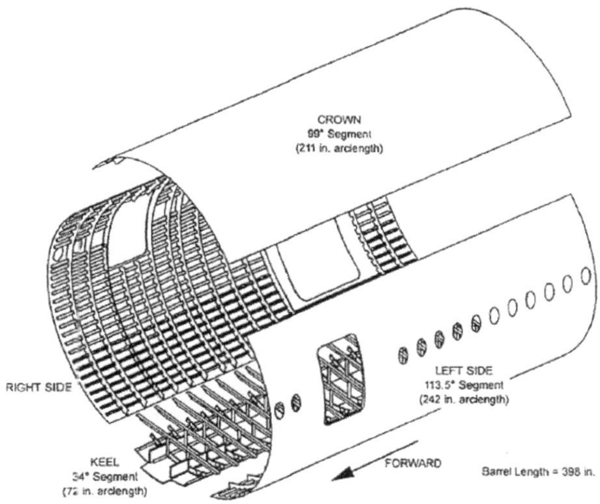

Figure 11. ACT quadrant panels [3].

The mechanically fastened frame approach greatly reduces the complexity of IML tooling needed to cocure the hat stiffeners and cobond the frames. This is especially true at the intersections of the frame and hat. Flexible caul plates and custom fit reusable bags became part of the tooling system needed to accomplish the fully integrated skin/stringer/frame structure. Producibility issues are complicated by the blind nature of the IML of the skin being completely covered by flexible cauls and the reusable bagging system. The structural arrangement shown in Figure 12 is very similar to the configurations that ended up on both the 787 and A350 programs.

The program studied the pultrusion process for producing skin stringers. Continuous resin transfer molding (CRTM) developed by Ciba-Geigy was one of the more promising technologies studied. Improved process control and reduced waste are among the perceived advantages; process maturity, constant cross-section stringers and costs associated with secondary bonding or cobonding are among the disadvantages.

Table 1.ACT structural concepts [3].

Details	Process
Skins	AFP (tow, hybrid AS4/S2) CTLM (contoured tape lamination machine, 12″ tape)
Frames	Braiding/resin transfer molding (triaxial 2-D braid) Compression molding Stretch forming (thermoplastic, discontinuous fibers) Pultrusion/pull forming
Stringers	Hat—ATLM/drape forming (cocured, thickness variation) "J"—pultrusion
Panel assembly	Cocured/cobonded stringers, cobonded frames Cocured/cobonded stringers, fastened frames Sandwich panels, cobonded frames

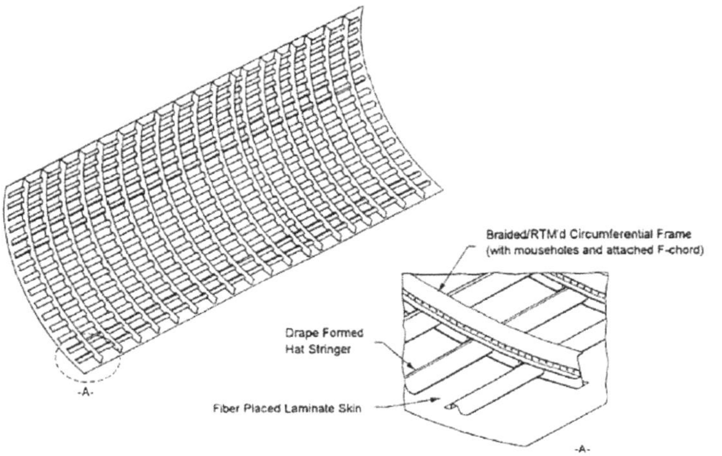

Figure 12. ACT crown panel structural arrangement [3].

Airbus has studied automating stringer fabrication using both pultrusion and RTM but felt limited by aspects of both processes. As an answer, Airbus developed their version of pultrusion RTM. Figure 13 shows equipment completed in 2011 that is being used to develop and qualify the process [4]. This hybrid fabrication approach allows the use of preform laminates instead of being limited to unidirectional reinforcements like traditional pultrusion and supports continuous production instead of batch processing associated with the traditional RTM. Instead of dipping the preform stack through a resin bath, it is pulled into an RTM tool that is open on both ends. To overcome resin being pushed out at both ends of the open tool, Airbus worked with resin suppliers to develop an epoxy resin with a parabolic temperature/viscosity curve. At 120°C resin viscosity is very low with high flow characteristics, but at both room temperature and at 180°C and higher, it is very viscous. The tool entry is cooled so the resin is too viscous to flow out; the middle is heated to obtain resin flow and cure; more heat is added at the end to increase resin viscosity to make sure it does not flow out and reduce cure pressure.

Figure 13. Airbus continuous pultrusion equipment [4]. Source: CTC Stade.

2.3. Automated Fiber Placement

Even in the early days of development, industry leaders believed in the possibility of higher layup rates using AFP than was possible with hand layup, but the capabilities and the scale that the industry has achieved today is astounding. Almost as astounding as how the industry reinvented itself from a raw material cost saving technology to an enabling technology for large aircraft structural components.

In the late 1980s and early 1990s Northrop and ATK/Hercules worked on several joint projects sponsored by the Air Force which included fiber placement development and application. The technology was in its infancy as ATK was developing tow placement (as it was more commonly referred to originally) from its roots in filament winding technology. The main prize in the early days was $5 per lb. high modulus carbon fiber and $15 per pound high temperature/high performance resin instead of the $60+ per pound price of prepreg. A wet process of running fiber through a resin bath prior to placement onto the layup mandrel was never able to realize the quality and consistency required by the design. This same process has been used in the large wind blade manufacturing process and it reminds us of how challenging (and messy!) that approach can be. In addition, the wind blade manufacturing industry has learned some valuable lessons from those early days of "build it as cheap as you can" using the lowest cost material you can deal with. While those early blades were built with lower manufacturing costs, the argument can be made that many of those blades failed very early in their lifecycle and required costly repairs or replacement to generate electricity. If the blade cannot turn because it has delaminated, it is not generating any electricity in addition to the cost of repair or replacement.

Not only did the technology not realize the cost savings of dry fiber and wet resin, it was forced to adopt prepreg technology into the process—namely dealing with backing paper and ADDING to the cost of unidirectional prepreg tape by requiring it to be slit into prepreg tows. At the time of the ATCAS program, the AFP process was still evolving from what was originally envisioned as a much lower raw material cost build up starting with a dry fiber/wet resin process instead of a costly unidirectional fiber prepreg. The baseline process the ATCAS program selected for fabricating fuselage skins was AFP using prepreg tow. The dry fiber/wet resin tow had evolved to prepreg tow in an attempt to improve process consistency. The process was selected based on several factors including the potential for reduced material cost (compared to prepreg tape), the potential to achieve high lay-up rates over contoured surfaces, and the potential to efficiently support a significant amount of ply tailoring. In addition, the fact that tow material does not require backing paper eliminated a perceived risk of greater machine downtime.

When compared with the quality and consistency of parts made with prepreg tape, tow preg and subsequent prepreg tow, was not acceptable. The variability seen in the quality of the resultant panels would require compensation in the design of the part, resulting in weight penalties. But this did not prove fatal to the technology, instead tow placement reinvented itself (Figure 14).

There have been many studies of the AFP process that have helped to shape and refine the characteristics and capabilities that exist in today's equipment offerings. But the ACT program allowed Boeing to better understand, study, define and refine the process to guide the technology development based on the needs of the user community. Everything from tack of the initial plies to the tool surface, to overlaps and gaps in the laminate; the most efficient ways to handle window/door cutouts, laminate thickness transitions, lay-up rates for flat, curved, cylindrical and duct shaped parts, etc., etc. What has ended up on production on the 787 is not the direct result of that ACT program, but the ACT program created the path for subsequent AFP development to follow and improve upon.

Figure 14. AFP process and tooling.

2.4. Tooling

One clear thread throughout the development of composite fuselage fabrication processes that was recognized and considered very early on, was tooling. The fabrication of large composite fuselage structures was also enabled by the tooling required to support it. The ability of industry to produce tools using specified materials and built to the size, scale and accuracy required by aerospace and defense applications were critical factors. Large scale machining, laser measuring systems, and innovative thinking supported the transition to today's composite fuselage manufacturing capability.

The ACT program demonstrated how the producibility of large, integrated, composite fuselage structures depend heavily on the tooling to ensure compatibility of the skin cure tool, the cocured or cobonded stringer tooling and the frame tooling. Controlling these elements is necessary to minimize gaps and interference fit between cured detail components. Understanding the effect of tolerance accumulations, warpage, liquid and hard shim allowances and fastener pull-up forces creates the ability to calculate the impact on fuselage structural arrangement and weight, part manufacturing cost and risk and fuselage assembly and integration time. These elements become even more critical as the size of the fuselage grows to 787 and A350 proportions.

One important note was the need for the stringer tooling to be extractable after cure and flexible enough to be able to accommodate skin thickness variations—especially the "joggles" or transitions up-across-down at each of the frame stations. These requirements drove the team toward silicone or flexible laminate mandrels—reusability was also a key consideration. The mandrels needed to be rigid enough for handling or to be used as drape or vacuum forming mandrels; durable and capable of withstanding a 350°F autoclave cure cycle and still be able to conform to skin ply sculpting and tailoring; and be able to be extracted after cure.

While the use of silicone mandrels and the flexible IML tooling proved adequate for controlling hat stiffener shape, quality and location for the demonstration panels, it was also recognized that silicone mandrels presented many challenges in both scale-up and production scenarios. Boeing started to develop hat shaped silicone bladders that fed autoclave pressure into the bladder throughout the cure to provide uniform pressure throughout the stringer. After cure, pressure in the bladder is released making it possible to remove the bladder.

At this same time Rubbercraft was working with engineers on the C-17 program to develop and manufacture inflatable silicone bladders for use on the replacement composite tail (Figure 15). In 1991 on aircraft 51, a composite tail was integrated into the program. Rubbercraft produced silicone bladders with FEP film molded to the OML of the bladders that were used in IML tools to cocure hat stiffeners to the skin of the horizontal stabilizers. The tooling, bladders and hat stiffener design allowed for the bladders to be manufactured with substantial excess length that supported multiple cure cycles despite the dimensional shrinkage of the bladder in the longitudinal direction. The reusability over multiple cure cycles is key to the cost effectiveness of the inflatable bladder system. Rubbercraft product improvement was focused on bladder attributes that supported increasing the number of cure cycles the bladder could be used for (Figure 16).

While Boeing was developing flexible IML tooling for cocuring hat stringers and cobonding frames on the ACT program, they evolved away from one-piece overall cauls to separate, individual flexible cauls constructed from graphite/epoxy fabric with a layer of Viton® fluoroelastomer and an outer layer of FEP film. The fluoroelastomer was shown to be more resistant to the epoxy resin and thus more durable than silicones or other rubbers. An added benefit—but perhaps not as well understood at the time—is the added resistance to permeability offered by both the FEP film and the Viton rubber. This helps to minimize the amount of autoclave gas on the inside the bladder from being introduced into the laminate through the permeability of the bladder system. Fluoroelastomer bladder development continues today in support of new programs and applications.

A comparison of OML and IML cure tool approaches demonstrates some of the tradeoffs that must be considered. OML tooling is less complex, less expensive, can be initiated as soon as the OML of the aircraft is established and is more forgiving of change than an IML tool. The IML tool requires less labor and risk for locating and maintaining locations of stiffeners and other elements and is much more simple to bag (Figures 17–20).

The ACT program also looked at separate male winding mandrels for AFP and then transferring the uncured skin to an OML cure tool. The male layup mandrel improved layup rates and proved to be a less expensive approach to meet production rate than two cure tools. This also plays to the argument for a combined IML controlled layup mandrel and cure tool—as Boeing selected for the 787 program.

Figure 15. C-17 horizontal stabilizer.

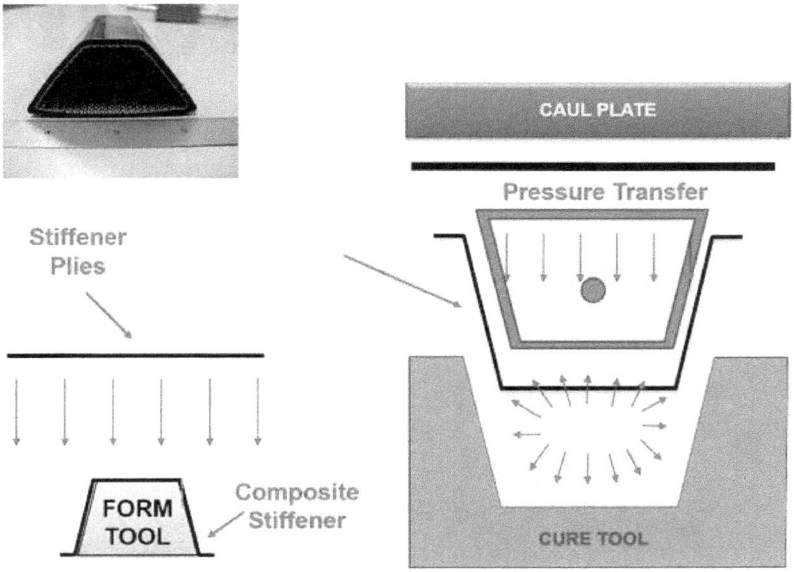

Figure 16. Inflatable bladder.

Figure 17. OML sector panel tool. Source: Premium Aerotec.

Figure 18. IML tool. Source: Boeing.

One concern using IML controlled cure tooling is the ability to adequately control the aerodynamic shell of the fuselage. For the ACT program this meant meeting surface waviness criteria of ±0.025″ over a 2″ length using caul plates.

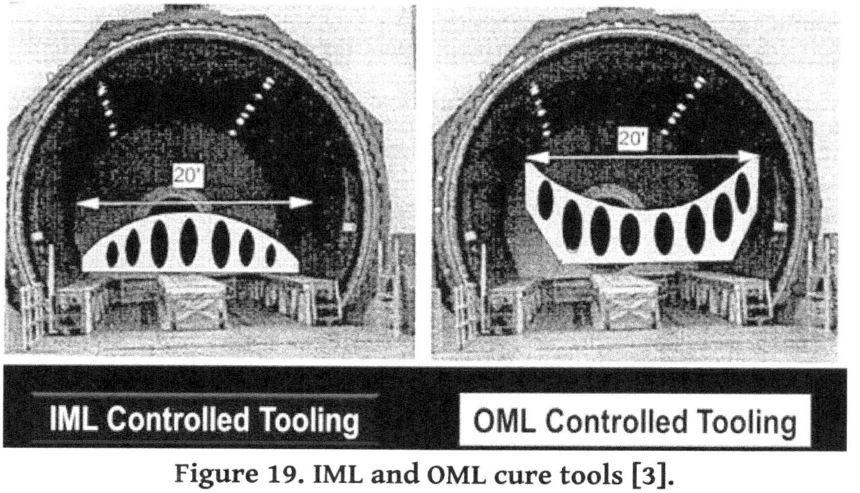

Figure 19. IML and OML cure tools [3].

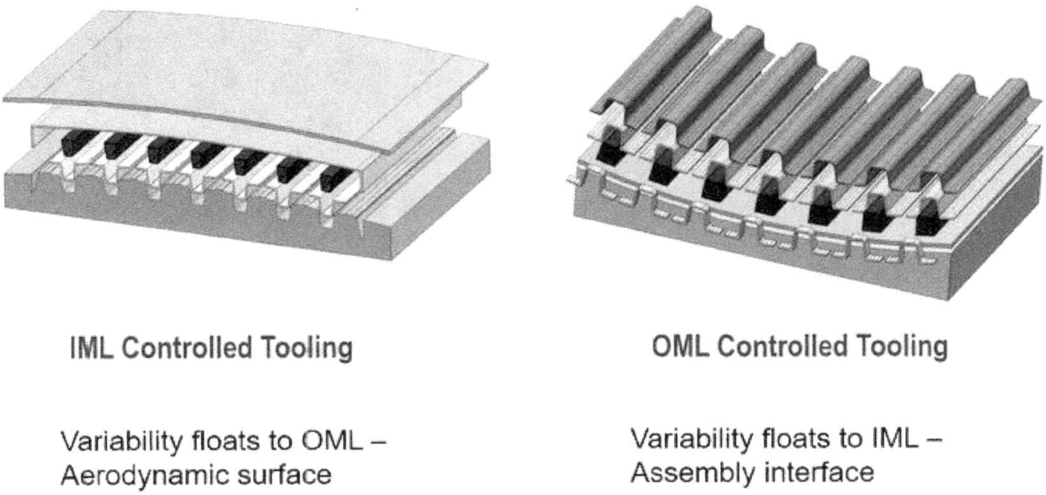

Figure 20. IML and OML tooling.

The concern over aerodynamic surface control seems to be greatly diminished when you look at what has evolved on the 787 program. The recognition that every airplane has a slightly different OML based on a number of factors such as exact resin content percentage in the prepreg (within the nominal tolerance range of ±5%), the amount of resin bleed experienced during cure and the amount of cured material removed during the sanding, smoothing and preparation for painting process. The skin of a composite fuselage allows for greater tailoring of the skin thickness than is usually incorporated into a metal fuselage. At the base, the fuselage is skin is thicker because it carries more load related to passengers, cargo and landing gear. The structural loads at the top of the fuselage are limited primarily to overhead bins, air ducting, and electrical wiring and this allows for lower weight, thinner skins that predominantly function as aerodynamic surfaces. Regardless of where in space it exists, and eventhough it varies from aircraft-to-aircraft, the surface is sanded smooth enough to satisfy the surface waviness allowance and negligible difference between aircraft.

The ATCAS team envisioned scenarios that included full one piece barrel fabrication. Significant cost savings were estimated from the elimination of longitudinal splices and the need to compensate for tolerance accumulation in assembly. Material out-time, segmented full barrel cure tooling and barrel warpage were the primary risks identified with full scale single piece barrel fabrication.

The sector panel construction used on the A350 allows for the use of invar for all the fuselage tooling. This includes the IML controlled sector panels fabricated by Spirit for Section 15. The approach Spirit applied is very similar to the one used on the 787 with the exception of the use of sector panels instead of a one piece barrel breakdown mandrel (Figure 21).

Figure 21. Tooling. Source: Boeing, Coast composites.

2.5. Large Autoclaves

One enabling capability that supports the evolution of the current state-of-practice for composite fuselage manufacturing is large autoclaves. There are many, many, many, many research and historical, ongoing and planned for the future, development efforts focused on OOA (or non-autoclave as it was called in the 1980s) materials and processes with the goal of eliminating that monument, the autoclave. The goal is noble (and not new) and the development efforts are making great progress and will, someday in the future, represent a significant (if not all) portion of the composite structure on commercial passenger aircraft—just not today. We already see components made from liquid molding processes being used in specific applications and families of parts and components on aircraft like the 787 and A350, just not the primary fuselage panels and stringers—yet. The maturity, forgiving nature, and low risk of baseline autoclave cured systems made it an easy decision for programs like the 787 and A350 to progress knowing that it was just time and money required to build autoclaves large enough to meet the needs of the program. No new technology needed, just scale and incorporation of improvements being realized by the autoclave industry, such as control systems and operational efficiencies. Spirit even built their own liquid nitrogen generating plant onsite to service their large autoclaves (Figure 22).

2.6. NDE/I/T Technology

The use of composites for high performance applications requires the ability to identify and ultimately eliminate structural defects that occur during manufacture, assembly, service, or maintenance. The entire field of nondestructive evaluation

(NDE) has continued to develop and evolve in parallel to the growth of composite structure applications.

Figure 22. Autoclaves. Source: Spirit, DLR.

It is both an enabling technology and one that has been driven by the market and the need. NDE of composites is a mature technology and has been used successfully for many years, however, the composite structures of today and tomorrow have grown in both scale and complexity. New and improved nondestructive testing (NDT) methods and technologies are necessary to improve detection capabilities, meet growing inspection needs, and address future nondestructive inspection (NDI) requirements. NDT methods currently used in aerospace applications span a broad range of technologies, from the simple coin tap test to fully automated, computerized systems that can inspect very large parts (Figures 23 and 24).

Many of the newer NDI methods are "wide-area" inspection techniques, which enable more uniform and rapid coverage of a test surface which can improve productivity and minimize human error. Technical advances in both computing power and commercially available, multi-axis robots and/or gantry systems, now facilitate a new generation of scanning machines. Many of these systems use multiple end effector tools yielding improvements in inspection quality and productivity.

Ultrasound is the current NDE method of choice to inspect large fiber reinforced airframe structures. Over the last 2 decades, ultrasound scanning machines using conventional techniques have been employed by all airframe OEMs and their top tier suppliers to perform these inspections. A limitation of ultrasonic inspection can be the requirement to use a couplant between probe and test part. VACRS (variable automatic couplant and recovery system) has helped changed the way very large area ultrasonic inspections are done [6]. The VACRS system uses a lightweight couplant and delivery/recovery system that makes it possible to conduct a C-scan with large ultrasonic arrays without the large water requirements. It works with Boeing's mobile automated scanner (MAUS®) and other scanning systems on the market.

Shearography and thermography are relatively fast, non-contact methods that require no coupling or complex scanning equipment. Laser shearography was initially applied to aircraft structure in 1987 by Northrop Grumman on the B-2 bomber. Since that time, laser shearography has emerged as an advanced, high-speed, high-performance inspection method.

Coin tap	Low-cost	Large, near surface features only
Ultrasound	Excellent penetration Well-established stands and procedures	Point inspection or scanning required Requires contact, couplant
Radiography	Area inspection Excellent crack detection	Insensitive to many void or delimitations
Thermography	Area inspection	Operates in close proximity to aircraft Limited depth range
Shearography	Area inspection	Limited depth range Issue at edges and corners

Figure 23. NDI methods [5].

An enabler for more widespread use of bonded structure in commercial aircraft applications will be improvements in cost and capability related to quantification of real-time structural bond integrity. Adhesive bonds degrade slowly over time and are highly dependent on surface preparation. On older aircraft, the only gauge for bond integrity is age, environmental exposures and statistics — not the actual condition of bonds. The ability to detect weak adhesive bonds, before they disbond will lead to more integration of parts and reduced fastener count and a reduction in everything that is involved with creating holes in cured composite parts. Military air vehicle platforms are more aggressive in this pursuit and the "pay-for-performance" mindset, the lower production rates and the size, visibility, and objectives of the programs allow for more flexibility in bonded structure implementation. The commercial world is different and just like the widespread implementation of composite material on new aircraft, it will not happen unless there are compelling economic advantages and very low risk.

Figure 24. Ultrasonic inspection.

2.7. Logistics

Boeing knew that the transport time required by land or marine shipping methods would not support a supply chain that included major partners located in Japan, Korea and Italy and that air transport would be the primary shipping method [7]. The Dreamlifter started as the Large Cargo Freighter (LCF) program and is a modified 747-400 freighter.

The Dreamlifter and follows a historic trail of oversized or outsize aircraft, which includes the Airbus Beluga, that were borne out of the adage "necessity breeds invention". The Dreamlifter is a dedicated transport used to deliver full 787 fuselage sections, wings, and horizontal tail from suppliers located across the US and the world. There are four Dreamlifters in operation supporting the 787 program.

The innovation that was the Dreamlifter (Figure 25), also required equipment to support the loading and unloading of such large cargo. Hence was born the largest cargo loaders in the world. The first one designated DBL-100 (DBL has been reported as an acronym for "Damn Big Loader"), were designed for use exclusively with the Dreamlifter.

Airbus was originally a consortium formed by British, French, German, and Spanish aerospace companies. Historically, each of the Airbus partners makes an entire aircraft section, which would then be transported to a central location for final assembly—even after integration into a single company, the arrangement remained largely the same. When Airbus started in 1970, road vehicles were initially used for the movement of components and sections. As production volume grew quickly, a switch to air transport was required. Beginning in 1972, a fleet of four highly modified "Super Guppies" took over. These were former Boeing Stratocruisers from the 1940s that had been converted with custom fuselages and turbine engines. Airbus' use of the Super Guppies led to the jest that that every Airbus took its first flight on a Boeing [8].

Today this need is handled by the Airbus A300-600ST (Super Transporter) or Beluga (Figure 25). The Beluga is a modified version of the A300-600 airliner adapted to carry aircraft parts and oversized cargo. The official name was originally Super Transporter, but the name Beluga, a whale, gained popularity based on the appearance of the airplane and has been officially adopted. Interestingly, the Beluga cannot carry most fuselage parts of the A380, which are instead transported by ship and road.

Figure 25. Beluga and Dreamlifter [7]. Source: Boeing, Airbus.

Airbus has an updated design, The Beluga XL, based on the larger Airbus A330-200. Five aircraft are planned to be built as replacements for the existing aircraft and used primarily for A350 work. The Beluga XL is designed with the capacity to ship two A350 wings simultaneously [9].

2.8. 787 vs. A350

The Boeing 787 and the Airbus A350 aircraft share many similarities in size, configuration, manufacturing methods and mission (Figure 26). The primary difference between the composite fuselage structures of the two programs is the exclusive usage of IML controlled cure tooling and full barrel fabrication applied by Boeing and the sector panel approach selected by Airbus with a high percent incorporation of cobonded fuselage skin stiffeners. The true results of these decisions will not be known until more information can be collected about actual fabrication and assembly costs being realized by Boeing and Airbus.

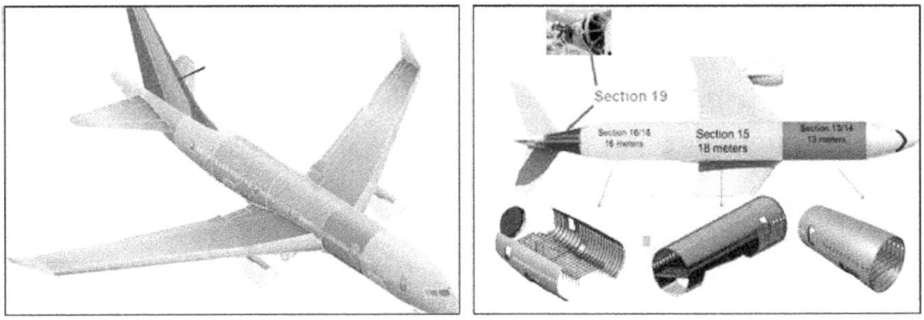

- One Piece Barrel Construction
- IML tooling

- One Piece Barrel Construction
- Sector Panel Construction
- Cocuring
- Cobonding
- IML tooling
- OML tooling

Figure 26. 787 and A350 fuselage sections.

2.8.1. Boeing 787

The ACT/ATCAS program had a tremendous influence on the direction Boeing selected for the 787 program. Lessons learned from all aspect of the program influenced everything from the material systems that were selected to the tooling materials, structural arrangement, and the selection of IML tooled, full barrel fuselage structures. Major considerations that influenced that decision were the concerns about the cost of the assembly of very large stiffened structure and the stresses induced on the structure due to assembly.

The program helped Boeing better understand the assembly loads related to composite panel warpage from cured part spring back and cocured and/or cobonded stiffener or frame mislocation. At minimum, these loads need to be understood and accounted for in the part design. Boeing saw an opportunity to minimize these assembly related penalties to the design by the tooling and structural arrangement approach applied on the 787.

Boeing's selection of the AFP process over a male mandrel that serves as both a layup and cure tool is forgiving enough to accommodate different caul plate approaches on different sections of the fuselage.

All the fuselage sections use multiple caul plates that nest together to cover the entire outer mold line of the fuselage. The cauls are floating on the surface of the skin and move with the skin during cure to establish the cured part OML whenever and wherever it is at the time the resin gels and things stop moving. Shared characteristics of the cauls include the ability to be individually and positively located before cure and removed individually after cure. Also the ability to ensure the cauls do not interfere with each other during cure. However, differences do exist in the choice of material (either graphite reinforced composite cauls or aluminum cauls) and in the thickness of the caul. In some cases, the composite caul is very thick and stiff and will behave more rigidly during the cure cycle. In other barrel sections, a thin aluminum caul is employed, which will more closely conform to the surface of the as AFP laminated skin. Both extremes are successfully being used by different fabrication partners.

Invar was the material of choice for Sections 43, 44 and 46 and the tail. Invar tooling was not the right choice for Spirit as it designed the layup mandrel/cure tooling for Section 41. An invar tool of that size and weight would have imposed very expensive requirements on the foundation of the AFP machine that winds the skin. The size of the motors and energy required to turn and manipulate the mandrel during the fiber placement process was also determined to be prohibitive. Instead Spirit elected to fabricate graphite reinforced BMI mandrels fabricated on invar cure tools and then machined to final IML dimensions (Figure 27).

Figure 27. Spirit 787 Section 41. Photo: Bill Carey.

Composite tooling is also used for Sections 47 and 48. In addition to lower mandrel weight, faster heat up and cool down rates contributed to this decision.

All the partners on the 787 program follow similar manufacturing processes for fabricating cocured, hat stiffened, full fuselage barrel sections. All use AFP over IML controlled male layup mandrels that also serve as cure tools. Each section (except the tail) uses multi-piece breakdown mandrels which are disassembled and removed from inside the fuselage after cure (Figures 28 and 29).

Figure 28. 787 Section 43. Source: Boeing.

Figure 29. Sections 44 and 46. Source: Boeing.

Alenia manufactures Sections 44 and 46 of the 787. Section 44 is a composite half barrel section that covers the main wing box. The lower portion of this fuselage section is mostly metallic and the structure is designed to handle the primary loads from the wings and landing gear.

Fabrication of fuselage barrel Sections 47 and 48 were originally contracted to Vought as part of their statement of work (SOW) on the 787 program. Financial pressures driven by initial program delays led to Boeing acquiring the Vought SOW including partnership in subassembly work with Alenia (Figures 30–32).

Figure 30. 787 Sections 47 and 48. Source: Boeing.

Figure 31. 787 Tail. Source: Boeing.

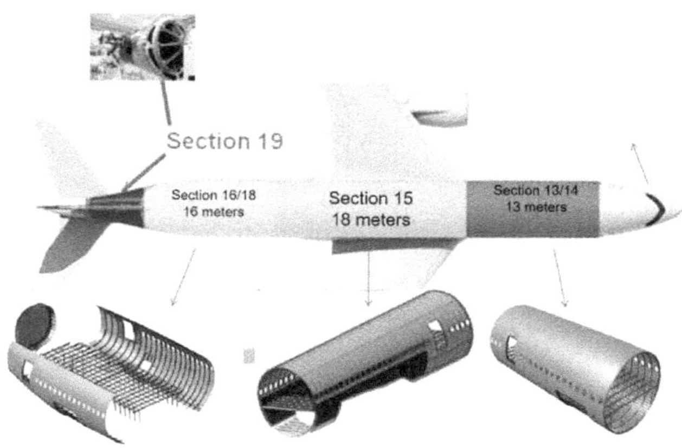

Figure 32. Airbus A350.

The tail is the only barrel section that does not require a breakdown cure mandrel. The natural draft angles allow for cured part removal by simply sliding the cured part off the mandrel.

Boeing achieved stretch version of the 787 by extending the fuselage sections on either side of the wing center of gravity. The 20′ stretch for the -9 was achieved by adding 10′ to Sections 43 and 47. The additional 18′ added for the -10 configurations was achieved by adding 10′ to the forward fuselage and 8′ aft end. When new AFP mandrels were added to meet production ramp-up rate needs and to meet the -9 configurations, the tools were designed to support -10 also.

2.8.2. Airbus A350 XWB

While the focus of this paper has concentrated on developments in the United States, the composites community in Europe was just as active.

There were many R&D programs that were directed at high performance composites design and manufacturing activities [10].

The results of this work along with many lessons learned on historical programs fed into the approach taken on the A350XWB program (XWB stands for eXtra Wide Body). The A350 composite fuselage manufacturing approach is not as uniform as the method selected by Boeing on the 787.

The A350 incorporates one complete barrel section, the tail, produced in Spain that uses an approach similar to the one used by Boeing and its partners on the 787 (Figure 33). The rest of the A350 fuselage follows a more conventional panel assembly approach, but with some unique manufacturing process used along the way. The use of AFP, invar tooling and longitudinally incorporated omega (like the Greek letter Ω) stiffeners, more traditionally called hat stiffeners, are also common between the programs. The panel approach used on the A350 supports long part lengths and this is reflected in Section 15 which is approximately 65′ in length. How the omega stiffeners are incorporated on the fuselage panels is quite different between sections and suppliers.

Spirit is a common key supplier on both programs and the fabrication approaches share some key characteristics. Spirit produces Section 15 of the A350 and applies the sector panel approach that is common throughout the fuselage. Spirit cocures the omegas using an IML controlled layup/cure tool with a stiff composite caul plate to control the aerodynamic OML surface smoothness. Uncured omega stiffeners are laminated, formed and located into troughs machined into the invar tool. Inflatable rubber bladders are located on top of the omega laminates and fill the void between the omega and the AFP skin that is laminated on top of over the assembly. The part is autoclave cured and the rubber bladders removed after cure leaving the cocured, and now hollow, omega on the panel (Figure 34).

The rest of the A350 fuselage structure uses cobonding to incorporate the omega stiffeners with the fuselage skin (Figure 35). Precured omega stiffeners are located onto green AFP skins with a layer of film adhesive between the elements and then autoclave cured (Figure 36). During the cobonding cycle shaped tube bags are located inside the cured stiffener and are open to autoclave pressure during the cure/cobonding cycle to ensure the already cured stringer does not collapse or become damaged when subjected to autoclave pressure (Figures 36 and 37).

Figure 33. A350 fuselage panel and tail. Source: Airbus.

Figure 34.A350 fuselage side panel. Source: Spirit.

Like the 787 program, liquid molding processes are used to fabricate fuselage frames which are mechanically attached to the skins. The structural arrangements and assembly methods used by both programs are remarkably similar.

One significant difference (if not THE most significant difference) is the frame integration to the fuselage. The 787 incorporates a "mouse hole" in the frame that nests around the hat stiffener and is attached directly to the IML of the fuselage skin. Boeing can do this because the IML surface of the 787 is a tooled surface with features that have controlled heights and locations. This includes hat stiffeners and skin joggles. Both programs use fuselage frames produced using a closed molding process that tools the surface that mates with the skin.

Figure 35. A350 fuselage panel. Source: CTC Stade.

Figure 36. A350 precured omega stringers. Source: Deseret News, Jeffrey D. Allred; CW/Photos: Jeff Sloan.

On the 787, this creates a tooled surface-to-tooled surface interface creating a very predictable assembly. Components fit together as well as it can be produced because early in the program, it paid the price of being designed for assembly

(Figure 38).

The A350 fuselage frames are attached only at the crowns of the omega stiffeners using secondary clips. Airbus did not try to attach the frames directly to the skins because the IML of the fuselage skin is not a controlled surface. It is a bagged surface that might use caul plates to create uniform pressure and a smooth surface, but the IML surface "floats" depending on factors such as bagging, resin bleed and initial prepreg resin content. Just as the OML of each 787 fuselage "floats" and is different aircraft-to-aircraft depending on these same factors. Airbus uses a standard carbon fiber reinforced clip, molded from thermoplastic material, to absorb the skin fabrication tolerance in the assembly process (Figure 39).

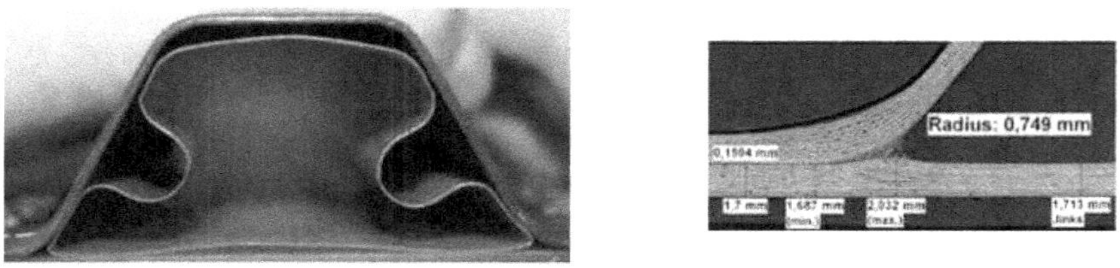

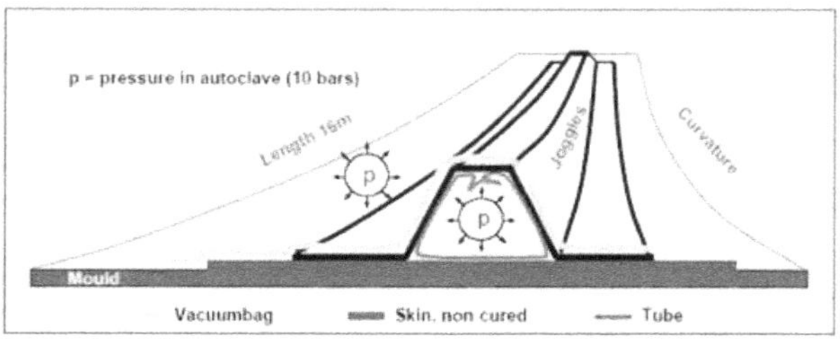

Figure 37. A350 omega stringer cobonding [11].

Figure 38. 787 fuselage.

Figure 39. A350 fuselage. Source: Borga Paquito.

3. FUTURE DEVELOPMENTS/TRENDS

There are several recently developed commercial aircraft, such as the Bombardier C Series, Mitsubishi's MRJ, and Comac's C919, that all have similar overall airframe architecture as the 787 and the A350. However, none of these aircraft incorporate an all composite fuselage. The advantages for composites on large, wide body aircraft have been validated by the short service history of the 787 and even shorter history of the A350. The debate regarding smaller aircraft achieving the same gains continues for Next Generation Single Aisles.

Wide body aircraft spend much of their life cruising at 40,000 ft. and the structure is sized for pressure loads and structural needs—this provides adequate thickness for good damage tolerant designs. The fuselage designs for single aisle aircraft could be more efficient based on cabin pressure and structural loading alone. But, to provide for designs that will be tolerant of many more takeoff and landings and in service hazards such as luggage and catering carts, dropped tools and equipment, hail and bird strikes, the fuselage panels must be thicker and heavier, thus sacrificing weight.

Wings are one area of implementation for composites on the single aisle upgrades and new aircraft of the future. The Boeing 777X has incorporated a composite wing into the design. A composite wing allows for a very high degree of laminate tailoring and can be designed and built for maximum efficiency. This creates an elegant wing that is incredible to watch in-flight, but appears alarmingly thin compared to conventional metal aircraft wings. But composite wings for high rates present challenges. Production rates of 12–14 per month for wide bodies have proven to be achievable. Building composite wings to support production rates as high as 60 aircraft per month for narrow bodies has not. Costs related to rate tooling alone can be daunting.

Remarkable advances in OOA technology might help provide a solution. Bombardier chose an OOA process for wings of the C-series and the MRJ is using an OOA system for the vertical tail wing box, a similar process to what United Aircraft (Russia) has announced for their MS-21 wing. Still, there are complex issues to resolve that will affect the timeline for OOA system usage on next generation, commercial, single aisle aircraft wings and fuselages. The industry is risk adverse and OOA systems are in their infancy compared to autoclave systems. The autoclave process has proven to be very forgiving and tolerant of variabilities that exist in raw materials, support materials, supply chain manufacturing processes and through final part fabrication. The effect of manufacturing variability is well understood and incorporated into efficient designs that contain minimal penalties for the unknown or less well understood. The same will not be true of OOA systems until more lessons learned have been earned. Many of these lessons will continue to come from military applications that are more aggressive in implementing new technologies. The benefit for the military is usually not cost; the benefit for the commercial world is always cost.

On a little longer timeline affecting future composite fuselage construction is sensor and technologies related to structural health monitoring (SHM). This is a very large field with growing interest by many OEM's in many applications by many industries, including aerospace, automotive, and power generation. Advances in this technical arena could be one of the next revolutionary changes or "step changes" (vs. evolutionary) to advance the industry. Advanced sensor technology could supplant many NDT applications by supporting in-situ "structural health monitoring." Installed on or within composite structures, such systems would continuously monitor a component and detects degradation and damage as it occurs. This could eliminate the possibility of damage being overlooked and reduce costly downtime for manual inspections.

The future of SHM and other smart composite structures includes morphing technology that changes part shape in-flight to create optimal flight conditions. Built-in sensing, computing, and actuation are emerging new frontiers for structures that self-tailor their properties for changing flight conditions. Similar developments include multi-functional composites—laminates that not only provide lightweight, load-bearing structures, but also perform additional functions such as energy harvesting and storage. The 20th International Conference on

Composite Materials (July 19–24, 2015, Copenhagen, Denmark), featured more than 100 presentations on multifunctional composites [12].

3-D printing is another emerging technology that will impact the future of composite fuselage construction. Already making an impact in prototyping, early design and development, and tooling applications. Small, highly complex parts will follow the path being created by 3-D printed metallic parts. Larger applications are sure to follow. Nano technology may also develop as a viable standalone technology or perhaps integrated with 3-D printing. Remarkable innovations are surely on the horizon.

4. CONCISE SUMMARY

Advancements in dual aisle, wide body commercial aircraft fuselages have progressed from small aerodynamic composite fairings and flaps to whole composite fuselage constructions during the last generation. The process has been systematic and persistent, involving dedicated efforts from military, defense department, university, and other industry actors. Teams in the US and Europe have engaged in a global competition, resulting in a mutually beneficial outcome for both parties. Significant technological advancements were necessary in various areas, including tooling and transportation. Significant progress was needed to improve the cost competitiveness of manufacturing and assembling composite materials to be used in commercial aircraft platforms. Upcoming mid-market aircraft models from Europe and the United States will introduce the latest technologies for dual-aisle commercial airplanes. The industry is waiting for guidance on the long-awaited replacement designs for single-aisle aircraft, namely whether they will use composite fuselages.

5. AT A GLANCE

The chapter discusses essential technology and manufacturing advancements that have facilitated the development of the composite fuselage. The author discusses many development, military, and production initiatives that have impacted the current condition of commercial fuselage construction. This article discusses the current methods and technology utilized in the production of wide body aircraft fuselages, as well as the possible causes for their adoption. Questions concerning the future of composite fuselage are raised by reflecting on past experiences.

REFERENCES

1. Palucka T, Bensaude-Vincent B. Composites overview
2. Tenney D. Darrel R, Pipes D, Byron R, Johnston N. NASA composite materials development: Lessons learned and future challenges

3. Willden KS, Harris CG, Flynn BW, Gessel MG, et al. Advanced technology composite fuselage—Program overview. NASA Contractor Report. 1997;4735
4. CompositesWorld. February 2017
5. Shepard S, Kulowitch P. Automated, Rapid Non-Destructive Inspection (NDI) of Large Scale Composite Structures. Available from: www.thermalwave.com
6. Bredahl B. NDT Solutions Inc.: Embracing Innovation and Providing Solutions. 2013. Available from: www.manufacturing.net
7. Available from: https://en.wikipedia.org/wiki/Boeing_Dreamlifterhttps:/
8. Available from: https;//en.wikipedia.org/wiki/Airbus_Beluga
9. Platoni K. Big ideas: Megalifters prove you're never too fat to fly. Air & Space Magazine. 2008
10. Herbeck IL, Kindervater IC. Ein neues Designkonzept für einen CFK-Flugzeugrumpf. Werksroffkolloquium 2006—Wettbewerb der Werkstoffe
11. Available from: http://www.maschinenmarkt.vogel.de/schlauchkerntechnik-sorgt-fuer-qualitaetssicherheit-a-373139
12. CompositesWorld. February 2016

6 Chapter

ROBOTIC AUTONOMOUS SPACECRAFT MISSIONS: CASSINI MISSION TO SATURN

1. INTRODUCTION: AN OVERVIEW

Spacecraft launched beyond Earth's gravity, such as satellites in orbit or interplanetary vehicles, must be self-sufficient in handling challenging flight conditions to achieve their scientific goals, unlike traditional aircraft that can be serviced when they malfunction. Once in space, these spacecraft are unable to return to Earth for maintenance and must rely on self-sufficient systems designed to prevent issues caused by human error, the space environment, incorrect commands from the operations team, or the significant delay between ground commands and spacecraft reception. Spacecraft must travel through space independently, protecting themselves from many factors that may pose obstacles to their internal health and functionality. Following the launch, spacecraft instruments are usually activated, its systems set up, and subsystem components are checked to ensure they are functioning correctly. The propulsion system is used during the mission to direct the spacecraft, modifying its path to reach the desired scientific objectives. These target objectives usually involve orbiting, flying by, or landing a spacecraft on objects like asteroids, moons, or planets. Scientific instruments are usually carried on spacecraft to conduct various scientific activities during the trip. The Deep Space Network (DSN) radio telescope array enables the Spacecraft Operations Flight Support (SOFS) team to communicate with NASA spacecraft during their missions. Commands are sent to the spacecraft through a "uplink," while detailed information is received through a "downlink" telemetry stream, including data on the spacecraft's systems, collected science data, and its journey encounters. This chapter details the obstacles encountered by the Cassini/Huygens Mission-to-Saturn spacecraft, the required preparations, and the flying experiences along its 20-year voyage to Saturn through our solar system.

2. CASSINI MISSION SUMMARY

The Cassini-Huygens Mission-to-Saturn interplanetary spacecraft mission was the fourth spacecraft to visit the Saturnian system, but was the first spacecraft ever to be captured into orbit about Saturn. The Cassini Program was a joint mission between NASA, the European Space Agency (ESA), and the Italian Space Agency (ASI), plus several other participants. Launched on October 15, 1997, Cassini traveled to Saturn following a 6.7-year cruise, which was supported by four Venus, Earth, and Jupiter planet "gravity-assist" flybys (an energy exchange between the planet and the spacecraft that accelerates the vehicle, changing its direction and velocity; Figure 1 [1]).

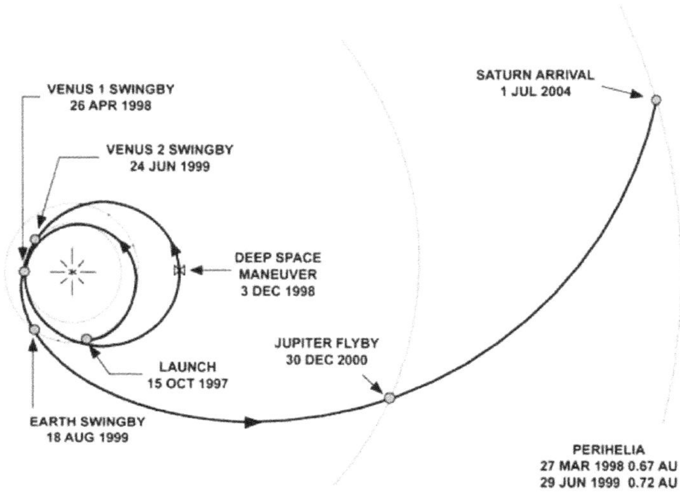

Figure 1. Cassini-Huygens mission trajectory to Saturn.

After a highly successful Saturn Orbit Insertion (SOI) in June of 2004, the Huygens Probe was deployed onto the surface of Titan in January of 2005.

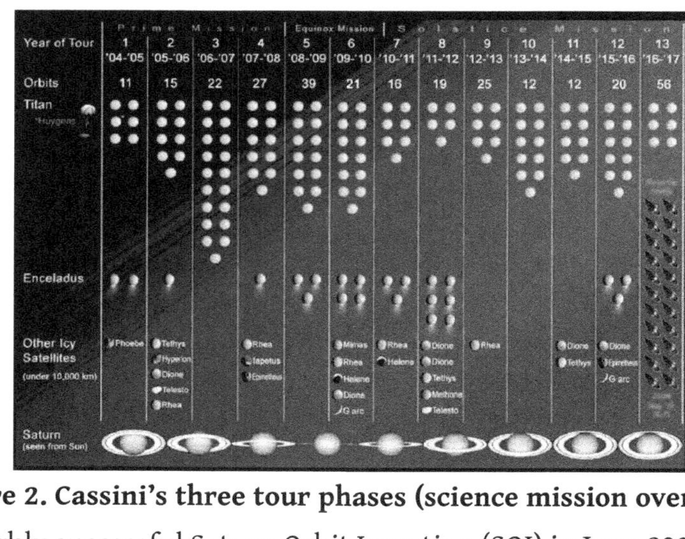

Figure 2. Cassini's three tour phases (science mission overview).

After a highly successful Saturn Orbit Insertion (SOI) in June 2004, the Huygens Probe was deployed onto the surface of Titan in January 2005. Cassini continued on to investigate Saturn, its rings, and satellites, in an outstanding 4-year expedition of the Saturnian system during its Prime Tour phase. At the end of the Prime mission (ending in 2008), all instruments and major spacecraft systems were verified to be healthy with a good volume of propellant remaining in the fuel tanks. Due to the great success of the Prime Tour with the vast quantity of new discoveries and overall quality of science returned by the spacecraft, NASA Headquarters allocated funding for the extension of Cassini's mission for a further 2 years, called the Equinox Mission (2008–2010).

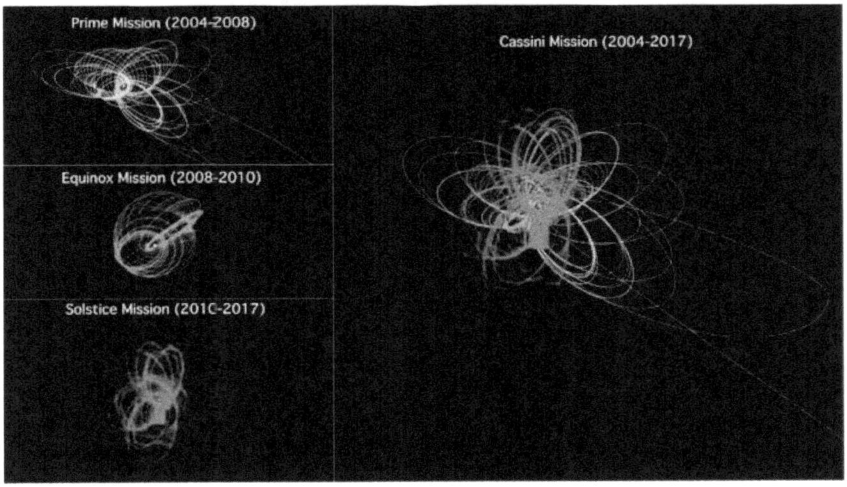

Figure 3. Cassini's orbital paths during its three mission tour phases (credit: NASA/JPL).

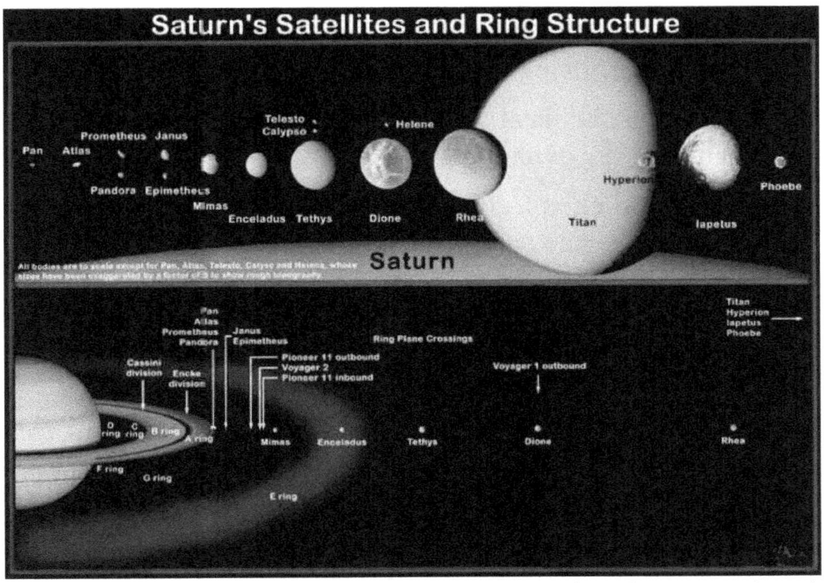

Figure 4. Saturn's ring structure and moon system (credit: NASA/JPL/Caltech).

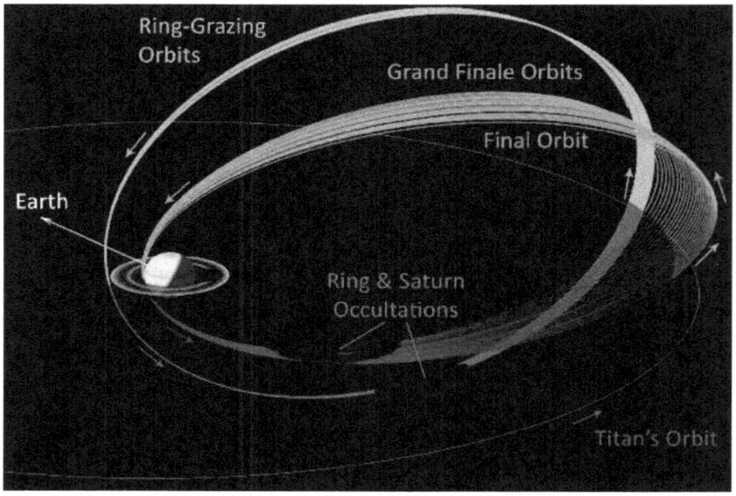

Figure 5. Cassini's proximal orbit phase & final plunge (credit: NASA/JPL).

A second mission extension was also granted, called the Solstice Mission (2010–2017) for a further 7 years of study (Figures 2 and 3). The Cassini mission ended with a final 42 orbit rotation through the outer and inner portion of the main ring system (F-Ring & D-Ring), followed by a fiery plunge into Saturn on September 15, 2017 (Figures 4 and 5, [2]).

3. BACKGROUND: SCIENCE GOALS AND CASSINI'S DESIGN

The science objectives of the prime Cassini mission were to determine:

- The elemental, molecular, isotopic, and mineralogical compositions of Saturn, Titan, the smaller moons, and Saturn's rings
- The physical, morphological, and geological nature of the above objects
- The physical and chemical processes within the atmospheres of Saturn and Titan (including the dynamics)
- The physical and chemical processes taking place on the surfaces of the rings and moons of the Saturnian system
- The physical and dynamic properties of the rings
- The composition and mapping of Saturn's magnetosphere and planet/ring interactions; interactions with the moons and solar wind
- The composition and mass distribution of ice and dust grains within the Saturnian system

And to:

- Map the surfaces of Titan and the icy satellites at wavelengths from extreme ultraviolet to Ku-band radar

- Measure plasma waves and radio emissions in the Saturnian system
- Examine the possibility of exobiology on Titan
- Search for gravitational waves

In order to achieve these science goals, several instruments were implemented onto the Cassini orbiter and Huygens probe vehicles. The combined Cassini-Huygens spacecraft consisted of 18 scientific instruments. Twelve instruments were placed on the Cassini orbiter (see Figure 6), and six were contained within the Huygens probe instruments (see Figure 7).

- Orbiter remote sensing instruments:
- Imaging Science Subsystem (ISS)
- Visible and Infrared Mapping Spectrometer (VIMS)
- Composite Infrared Spectrometer (CIRS)
- Ultra-Violet Imaging Spectrograph (UVIS)
- Radar
- Radio Science (RS)
- Orbiter fields, particles, and waves instruments:
- Dual Technique Magnetometer (MAG)
- Radio and Plasma Wave Science (RPWS)
- CAssini Plasma Spectrometer (CAPS)
- Magnetospheric IMaging Instrument (MIMI)
- Cosmic Dust Analyzer (CDA)
- Ion and Neutral Mass Spectrometer (INMS)

Huygens probe instruments:

- Huygens Atmospheric Structure Instrument (HASI)
- Aerosol Collector Pyrolyzer (ACP)
- Gas Chromatograph/Mass Spectrometer (GCMS)
- Descent Imager/Spectral Radiometer (DISR)
- Doppler Wind Experiment (DWE)
- Surface Science Package (SSP)

The instruments on the Cassini orbiter were body-mounted (no scanning platforms), which required the spacecraft to be oriented toward specific science targets for some instruments. Optical instruments provided imagery and spectrometry, while the Radar supplied imaging, altimetry, and radiometry. Radio links contributed information about intervening material and gravity fields. Other instruments on the orbiter were used to measure electromagnetic fields and the properties of plasma, energetic particles, and dust particles.

The Huygens probe was spin-stabilized, returning data via an S-band link to the Cassini orbiter. The probe's six instruments included several sensors to determine the atmospheric properties and composition of Titan. The probe's radiometric and optical sensors produced data on thermal balance and captured images of Titan's atmosphere and its surface. Wind profiles were captured by Doppler measurements between the probe and orbiter. Surface sensors on the probe were implemented to measure the surface impact acceleration, in addition to thermal and electrical properties.

The combined Cassini-Huygens instrument suite enabled scientists to determine the composition, physical, morphological, geological nature, and chemical processes of Saturn and Titan's atmospheres, to investigate their surfaces, and the magnetosphere of the Saturnian system.

3.1. Cassini Orbiter Subsystem Functional Descriptions

Cassini was a "stacked configuration" containing a lower equipment module, a propulsion module, an upper equipment module, and a High-gain Antenna (HGA). The Huygens Probe, Remote Sensing Pallet, and Fields & Particles Pallet of scientific instruments are attached to the stack within the upper equipment module, which contains the orbiter's 12-bay electronics bus, along with an 11-m magnetometer boom. Several engineering subsystems/devices control the spacecraft's operation as defined below:

Command and data subsystem (*CDS*): The CDS consists of two redundant computers that receive ground commands and memory loads through the RFS subsystem, processing and distributing the data to designated instruments and subsystems. CDS also receives data from Cassini's various subsystems and instruments, processing and formatting the data into telemetry packets by applying Reed-Solomon encoding, and then delivering the data to earth-based DSN ground stations through the telemetry data stream via the RFS subsystem. CDS also contains two Solid State Recorders (SSRs) with a 2.01-gigabit mass storage capability for Flight Software (FSW) loads and captured science data.

Attitude and articulation control subsystem (*AACS*): The AACS is comprised of two redundant computers which provide three-axis stabilization attitude control by either reaction wheel assembly (RWA) control or by the reaction control system (RCS) thrusters. Two sun sensor assemblies (SSA) and two stellar reference units (SRU) provide celestial attitude reference. Inertial reference is furnished by vibrating (nonrotating) gyros. An accelerometer on the central z-axis aids in controlling the duration of the engine burns. AACS flight computers receive commands from the CDS by way of a data bus, sending commands over its own data bus to the AACS controlled assemblies.

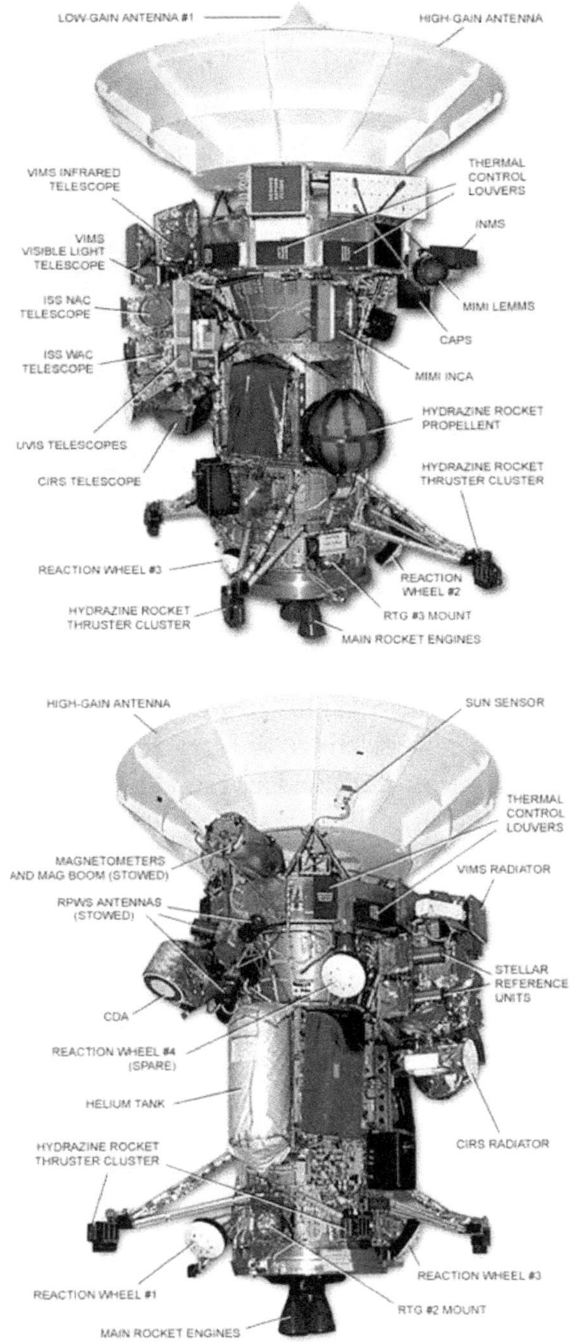

Figure 6. The Cassini orbiter instrument and device suite.

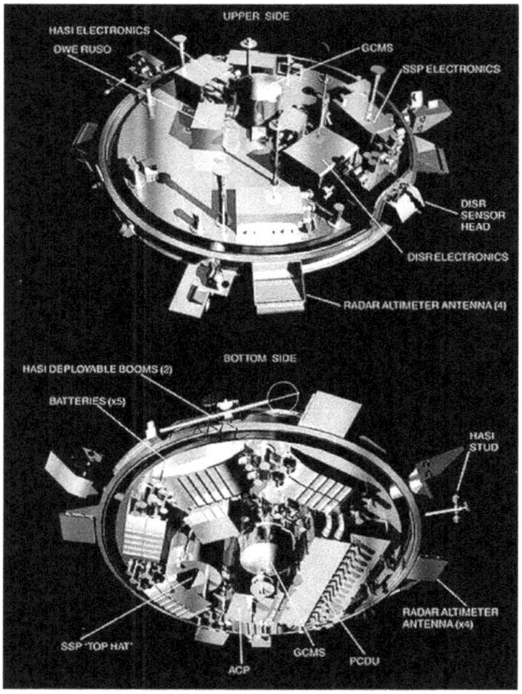

Figure 7. Huygens probe instrument suite.

Propulsion module subsystem (*PMS*): The PMS contains two redundant (gimbaled) 445-N engines with a specific impulse of 3020 N-s/kg (308 lbf-s/lbm), respectively. Approximately, 3000 kg of nitrogentetroxide and monomethylhydrazine are housed in the main bipropellent tanks. A retractable cover protects the main engines (ME) from damage by dust and micrometeoroid impacts. Four sets of mono-propellant hydrazine RCS thrusters (0.2–1.0-N thrust) fire in a direction parallel and perpendicular to the HGA (130 kg hydrazine tank capacity). Helium pressurization feeds the ME and RCS liquid propellants.

Power and pyrotechnic subsystem (*PPS*): Power was provided by three radioisotope thermoelectric generators (RTG). At the beginning of Cassini's mission, an allocation of 882 W of power was available, declining to 600 W by the end of the Solstice mission. The PPS distributes regulated 30-V dc power to orbiter instruments and subsystems by way of a power bus and 192 solid-state power switches (SSPS). Firing of pyrotechnic devices is supplied by the PPS once commands are received by the CDS subsystem. A shunt radiator disposes all unused heat energy from the RTGs by radiating the excess into space.

Radio frequency subsystem (*RFS*): An X-band link (7.2–8.4 GHz) provides communication between the ground and Cassini orbiter. Contained in this subsystem are redundant deep space transponders (DST; each includes a receiver and an exciter), and redundant traveling-wave tube power amplifiers which provide a 20

W radio frequency output [3]. The RFS also includes two command detector units (TCU), an ultra-stable oscillator (USO) for the radio science investigation, as well as an auxiliary oscillator. Telemetry modulation units and interface control units are also included. DSN station support is supplied through the use of 70- and 34-m ground antennas for uplink commanding, and to capture downlinked telemetry.

Cassini's antenna suite consists of a 4-m parabolic HGA and two low-gain antennas (LGA) fixed to the structure of the vehicle. Communication is accomplished through an X-band feed. To receive telemetry from the Huygens probe after Cassini/probe separation, an S-band feed was used. A Ka-feed and 5 Ku-feeds supplied additional beams for radar experiments.

Thermal subsystem: The thermal subsystem provides control of vehicle temperatures by the application of reflective multilayer insulating blankets, radiators, reflective and absorptive paints, louvers, shades, radioisotope heater units, and electrical heaters. For selected devices, autonomous thermal control (ATC) is applied.

In general, redundancy was applied to devices whose failure could cause loss of the mission, or loss of data from more than one scientific instrument. Onboard fault protection (FP) was designed into the system to safeguard against many possible fault conditions. Most electronic parts were radiation hardened and designed to be resistant to single-event upsets (SEU).

4. CASSINI'S MISSION CHALLENGES

Before spacecraft like Cassini can be launched, designers must consider external and internal influences on all devices and instruments. These components must be monitored, regulated, and controlled on a continuous basis during the entire lifetime of the mission.

4.1. Flight Environment

Temperature conditions internal and external to the spacecraft must be monitored constantly. The vacuum of space exposes the spacecraft to intense heat from the sun when the spacecraft is in close proximity, causing its surfaces to superheat. Shadowed surfaces are subject to extremely low temperatures which can cause onboard propellants to freeze. Once frozen, the spacecraft will be rendered inoperative, since it inhibits the spacecraft's ability to maneuver, so that it will eventually become misaligned with the earth (and unable to receive ground commands). Material stresses are also a concern with these temperature extremes, since thermal expansion-contraction can introduce camera distortion, breakage of components, and warpage. Also, computers and spacecraft components will cease to work if temperatures become too extreme. Instruments can fall out of operating limits, since many devices only function properly within a narrow range of temperatures. Heat build-up can also occur from the spacecraft's own systems. For Cassini, several protective measures were applied to control these hazardous conditions: the application of reflective multilayer insulating blankets to reflect the sun's heat, radiators were added, reflective/absorptive paints applied, louvers and shades installed, radioisotope heater units added, in addition to the inclusion of electrical heaters and ATC controlling monitors. Internal temperatures were also regulated by circulating the spacecraft's liquid fuel to cool its interior. When flying within the vicinity of the sun, Cassini shielded itself from overheating by utilizing the HGA as a sunshade.

Micrometeoroid bombardment, cosmic rays, and radiation are also part of the hazardous flight environment, having the potential to damage or interfere with the operation of the spacecraft's subsystems. Radiation-hardening was applied to electronic devices to deal with this risk, and thermal blankets and commanded HGA shielding of spacecraft components (in the direction of flight) was used to protect against micrometeoroid impacts.

Some unknown influences were also in play for the Cassini mission. The unique (and partially unknown) dust environment at Saturn, which can potentially influence component operation or become hazardous to the spacecraft during flight, would be a new and unique flight environment for the mission. Cassini was also the first JPL mission ever to use SSPS for power distribution, and its operation under these external influences could potentially be affected.

4.2. Planning for and Maintaining Consumables

All spacecraft must maintain adequate power margins to operate their subsystem components and scientific instruments, and to support communications with earth. Cassini's electrical power was derived from three RTGs, with a Beginning-of-Mission (BOM) capability of 875 W. RTGs are lightweight, compact power systems that are extraordinarily reliable. RTGs have no moving parts and provide power through the natural radioactive decay of Plutonium-238. The heat generated from the natural decay is converted into electricity by solid-state thermoelectric converters, enabling spacecraft to operate at significant distances from the sun, where solar power systems could be infeasible or ineffective compared to other power solutions. The durability and dependability of RTGs made them the preferred choice to implement the Cassini mission and its extended operation in the distant environment of Saturn orbit (~10 AU from the sun). The power output from the RTGs decreases predictably over time, so that the number of powered loads allowed to operate simultaneously must also decline accordingly. Planning and predicting the allowable number of operating spacecraft power loads (devices) is necessary throughout the mission as the available power decreases.

There are several other consumables which must be monitored on the spacecraft as well. The fuel and oxidizer used by the ME system (plus the hydrazine of the RCS) are particularly valuable, in that their availability controls the useful lifetime of the spacecraft. This is an important commodity for the consideration of mission extensions. Sufficient fuel for the end of a spacecraft's mission must also be maintained so that disposal of the vehicle is adhered to under planetary protection plan constraints [4].

4.3. Protecting Against Human Error

Human interaction with the spacecraft design and operation must also be considered when designing its systems against possibly fault occurrences. Human-induced error can manifest itself in the form of electro-static discharge events with spacecraft components during the manufacturing process. These are referred to as "latent failures" and can sometimes present themselves well after launch, rendering a device partially or completely useless. Commanded sequences that are uplinked to the spacecraft during mission operations contain instructions for data collection and control of spacecraft's activities, and can contain errors as well. These onboard running sequences (that execute continuously for weeks to months) consist of hundreds of commands to perform activities such as earth, sun, and star tracking, monitoring celestial references for attitude targeting, performing maneuvers to fine-tune the trajectory when required, science calibration and collection; all of which are all subject to human-induced glitches which can potentially cause serious faults. As an example, should the transmitter or receiver onboard the spacecraft be accidentally commanded off, the condition would cause an inability of the ground station to communicate with the spacecraft [5]. Too many components commanded on at the same time could exceed the spacecraft's power allocation, leading to a spacecraft-wide "under-voltage power-outage" condition. An error in target parameters could send the spacecraft in the wrong direction or miss a valuable science observation.

The possibility of human error must also be considered during the spacecraft's conceptual design process where prelaunch assumptions are made based upon past mission experience, in some cases, using their test data which is not an "apples-to-apples" comparison as assumed.

4.4. Aging Hardware

After many years of flight through the harsh flight environment, it is expected that spacecraft will experience various hardware degradations and failures. These potential problems must also be taken into account when extending spacecraft missions past their intended prime mission end dates, as the functionality of critical devices, is clearly a factor in this decision. Sensors can fail and devices that must undergo periodic cycling are all subject to breakdowns and degradations, which limit the mission's capability to perform future planned objectives.

4.5. Dealing with Earth-Spacecraft Relative Distance

An inhibitor of fault diagnosis and resolution is the ever-increasing lag time experienced on missions with large earth-to-spacecraft distance, referred to as Round Trip Light Time (RTLT). Ground ⇒ Spacecraft ⇒ ground transactions are almost instantaneous when the vehicle is near the earth since radio waves travel at the speed of light, but once the spacecraft gains substantial distance from our planet, even a signal traveling at this great velocity can take hours. In the case of Cassini at Saturn, a command sent from the ground took nearly 3 h to confirm back on Earth (~10 AU). This lag time becomes a high-risk deterrent to resolving problems when spacecraft like Cassini are sent out great distances. In fact, under certain failure conditions, it is impossible for the ground team to detect a spacecraft's anomalous condition and command recovery actions in time to preclude a catastrophic failure from occurring. An example of this situation would be failure of the helium latch valve to close properly (within the PMS system) after a pressurization task of the fuel/oxidizer tanks. This valve failure could cause the tank pressure to rise substantially in a very short period of time. If this condition occurred on the Cassini spacecraft, the pressure could rise to a catastrophic level before the pressure measurement data can even reach earth's ground stations to indicate the fault condition. In addition to fault detection and resolution concerns, this large lag time becomes a significant factor in the presence of one-time science opportunities such as planet flybys, moon encounters, and special science targets. For these events, the timing is crucial since only one opportunity exists to meet the objective and there may be no second chance. In many cases, these unique events must proceed unimpeded by fault interference in order for the spacecraft's mission to be successful.

4.6. Meeting Challenging Problems Through FP & FSW Uploads

To aid in many of the above challenges, onboard autonomous Fault Protection routines are implemented into the computers' FSW to monitor the spacecraft's many systems and devices to autonomously detect fault occurrences and respond to anomalous conditions. FP consists of "canned" automated responses that can swap to redundant devices (if available), command actions (like closing valves, commanding alternate targets, etc.) and/or place the spacecraft into a "safe state" using preprogrammed instructional routines. A general-purpose, "Safe Mode" fault response routine is typically executed if the fault condition interferes with the onboard running sequence (along with other corrective actions performed by FP if required), which terminates the onboard running sequence, configures the spacecraft to a lower power state by powering off all nonessential spacecraft loads, commands a thermally safe attitude and safe state for the hardware, establishes a low uplink and downlink rate for earth communications, and commands the LGA antenna (to accommodate the low rates). This safe, predictable spacecraft state allows the SOFS sufficient time to evaluate the fault causes and determine a solution [6]. On Cassini, FP was implemented early in the design phase. In general, FP responsibility is allocated to both the SOFS team and the spacecraft (which must deliver sufficient information on its health and fault condition to support fault recovery).

Unexpected conditions and problems can potentially exist for spacecraft missions that are exploring unknown parts of our solar system. New devices never flown in space before can experience unexpected faults due to the adverse flight environment. For these reasons (and those stated above), designers provide the SOFS team with the ability to upload FSW patches (replacing the memory locations within the onboard FSW with new data), and to replace entire CDS, AACS, or instrument FSW loads so that unknown problems can be addressed and increased visibility added to the downlinked telemetry stream.

5. CASSINI MISSION EXPERIENCE

Cassini-Huygens is a "Class A" Flagship mission, which requires that it be configured as a low risk, high robustness design with all practical measures taken to assure mission success. Numerous analyses and test programs were required before launch approval could be obtained for Cassini by NASA, in order to assure the mission's technical worthiness. These programs were also needed to fulfill mission requirements, which consisted of spacecraft loads analyses to demonstrate that all structural margins met expected safety standards, including a modal test program that yielded experimental data to verify the spacecraft and instruments via a finite element model arranged in the launch configuration. Dynamic tests of the spacecraft and instruments were also performed, as well as acoustic/vibration tests [7, 8, 9, 10]. Thermal analyses provided environmental verification, proving the functionality of all components. Also, verified were the heater power and the radiator area for engineering, as well as transducer performance [11, 12].

For Cassini, JPL's "conceptual life cycle strategy" was implemented. This consisted of splitting the development effort into several phases [13]:

- Prephase A: Advanced studies
- Phase A: Mission & systems definition
- Phase B: Preliminary design
- Phase C: Detailed design

- Phase D: Build & test
- Phase E: Operations

Once in Phase E, Cassini's operations phase was also divided into phases. Each phase was executed by way of several uplinked command sequences which were stored and executed onboard the spacecraft (sequences were designated as "C" for cruise or "S" for science):

- Launch and Deployment (C1–C4 sequences)
- Inner Cruise (C5–C16 sequences)
- Outer Cruise (C17–C32 sequences)
- Science Cruise Phase (C33–C41 sequences)
- Saturn Approach Science Phase (C42–C44 sequences)
- Saturn Tour (S01–S06 sequences)
- Huygens Probe Mission (S07 sequence)
- Tour (all three tours; S08–S101 sequences)

The Launch Phase spanned from launch (L) to L + 30 days, during which time launch activities and essential engineering checkouts and calibrations were required to prepare for the first main engine maneuver at L + 25 days. The Inner Cruise Phase encompassed the trajectory interior to Earth's orbit, included two Venus flybys and an Earth flyby. In this phase, the two close flybys of Venus and Earth were required to gain the needed velocity boost through gravity-assist maneuvers, to allow Cassini's trajectory to continue on to Saturn (via the next flyby at Jupiter). The science activities during this period were limited to instrument checkout exercises, with limited science performed during the Venus and Earth flybys. Since Cassini was in close proximity to the sun, the HGA was used to shield the spacecraft to prevent overheating.

During the Outer Cruise Phase, the HGA was used for data transmission (instead of the LGA) since the relative distance between Cassini and the sun was now increasing rapidly and overheating was no longer an issue. Instrument checkout activities continued during this phase, as well as checkout of the Huygens probe. Also included was the final gravity-assist flyby of Jupiter, where extensive science activities began. The Science Cruise Phase began 2 years prior to arrival at Saturn, in order to prepare for Cassini's arrival. Science activities increased during this time, and final instrument calibrations were completed.

The Saturn Approach Phase included a one-time opportunity flyby of the Phoebe moon and the SOI deceleration burn. After launch, the SOI burn was the most crucial activity of the entire mission since it not only allowed Cassini to be captured into Saturn's orbit, but also was an opportunity to view the planet at the closest range of the entire Prime Mission. The Probe Mission Phase was completed on the third encounter (flyby) with the moon Titan. The Tour Phase began at SOI and continued for 13 years (including the two extended missions, Equinox and Solstice). The moon Titan was massive enough to offer gravity-assist capability, and was used as "the tour engine" enabling orbit rotation, orbital period, and inclination changes needed to study Saturn's geometry, as well as to set up the many icy satellite encounters.

5.1. Prime Mission Experience

During Cassini's mission and its three tour phases, there were several instances where faults and problems occurred that required resolution by way of the onboard FP, FSW updates, and/or SOFS interaction. Detailed in the following sections are some of these experiences (mostly unexpected) during the Cassini mission, which challenged prelaunch assumptions and the ingenuity of the SOFS team.

5.1.1. Launch and Deployment

At Cape Canaveral, Florida, final preparations were nearly complete for Cassini's launch from Space Launch Complex 40 (SLC-40). But on September 3, 1997, NASA announced that high air conditioning flow-rate servicing of the Cassini spacecraft and the Huygens Probe tore a 2-inch rip within the insulation protecting the probe. It was feared that particles may have contaminated Huygens' delicate instruments, so that the spacecraft had to be hoisted off the launch tower, and the Huygens Probe removed and cleaned thoroughly. Re-installation of the probe on Cassini was performed on September 13 and the Cassini/Huygens vehicle was returned to SLC-40, followed by the integration of the spacecraft with the launch vehicle.

Cassini had a 30-day nominal launch window (from October 10, 1997 to November 4, 1997), which provided an arrival date at Saturn of January 7, 2004. After this launch window expired, the desired arrival date would no longer be achievable. A Titan IV launch vehicle with Solid Rocket Motor Upgrades (SRMUs) and a Centaur upper stage was used as the launch vehicle; Cassini was the second mission to use the SRMU configuration. Cassini was scheduled to launch on October 13, 1997, and after two launch attempts, the spacecraft successfully achieved lifted-off on October 15, 1997 at 08:55 UTC. Cassini was placed into an elliptical orbit by the Centaur upper stage burn (170 × 445 km parking orbit with an inclination of approximately 30°). In case the Centaur stage failed to successfully initiate a successful second burn, this "parking orbit" was designed to provide an orbital lifetime of about 20 days. Failure of subsequent burns would have caused the SOFS team to initiate operations to keep the spacecraft in a Sufficiently High Orbit (SHO) so that Cassini could be placed into a 2000-year lifetime orbit. But after 17 min in the parking orbit, the Centaur successfully fired again, launching Cassini toward Venus en route to Saturn. Cassini's AACS computers then executed the "find stars" mode block to acquire star knowledge via the onboard sequence, starting its journey toward the Saturnian system.

SSR bit flips: Almost as soon as Cassini left the launch pad, the spacecraft's telemetry stream indicated a higher than expected single bit error (SBE) and double bit error (DBE) rate in the SSRs than was predicted by the SSR Specification document. This spec predicted occurrences of SBE = 6/week and DBE = 2/year per SSR; the actual in flight was SBE = 20/h and DBE = 2/day. The SSRs are a high capacity, solid state bulk storage medium with no moving parts, containing 2.01 gigabits of memory per SSR for storage of computer/instrument FSW and collected science data. These erroneous "bit flips" change the affected stored/collected data from "1" to "0" (or vice versa), corrupting the data. Error detection and correction (EDAC) logic was installed by the manufacturer to "scrub" (detect and fix) the SBEs every several minutes, but the DBEs cannot be corrected without an arduous manual process performed by the SOFS team. An anomaly team was formed to determine the cause for these high bit rates. The team discovered that due to the physical adjacency of some data and checksum bits (a violation of design requirements), one cosmic ray could cause two bit errors to occur [14]. This was due to a human error in the mapping of SSR memory.

Fix: As a result, a new "SSR DBE Auto Repair" FP algorithm was designed by the SOFS team and uplinked to detect and initiate automatic repairs of DBEs within the FSW on both SSRs.

PMS regulator malfunction: The spacecraft prepared for the first Trajectory Control Maneuver (TCM) on November 9, 1997 (L + 25 days). Before this first maneuver could begin, the fuel and oxidizer tanks were heated (in order to avoid an irreversible overpressure in the propellant lines), including venting, priming, and pressurizing of the bipropellant lines for the ME. This venting activity removes the gas between the latch valves and the engines, which creates a vacuum in the propellant lines. The ME cover was opened prior to venting, and the lines were primed (priming fills the ME lines with propellant). The helium pressurant line was opened (to fill the ullage bubble within the fuel and oxidizer tanks) by opening a pyro valve, PV-1.

However when PV-1 was opened, the prime regulator (which keeps the tank pressures at a safe level) was discovered to have malfunctioned due to a trapped particle within the hard-seat regulator, and was leaking at a significant rate. The pressure in the tanks rose high enough to reach FP thresholds, which would have activated the Over-pressure Response FP, executing the Safe Mode Response and halting the onboard sequence (and the ME burn maneuver). Analysis determined that the leak rate was 1700 cc/min; the worst leak rate expected through testing was only 1.70 cc/min (a factor of 1000 times lower than this leak rate). The impact of this unexpected regulator malfunction would now require a substantial redesign in the ME burn strategy for the entire mission. This leak further increased a year later during the 90-min Deep Space Maneuver (DSM) burn, by a factor of 6.6. The upcoming SOI burn (in the next 6 years) was a crucial mission event which relied upon the characterization of the PMS system 30 days before Saturn-capture. This task would now be impossible to achieve, so that an entire redesign of the 90-min SOI burn would now be required [15].

Fix: To halt the pressure rise, the SOFS team uplinked a command to close the High Pressure Latch Valve (HPLV) to stop the helium pressurant from filling the tanks' ullage bubble with helium. During the cruise period, the mission was redesigned so that all ME burns were supported by a special uplinked sequence which controlled the inflow pressurization of the fuel and oxidizer tank duration, by allowing the HPLV to remain open for just a short period of time (~10 min). A new set of FP routines addressing the associated new failure modes that resulted from the redesign effort were also uplinked to the spacecraft's FSW, and the SOI burn pressurization strategy was also redesigned successfully.

5.1.2. Inner Cruise

Safe mode activation #1: FP swapped the prime SRU to the backup device during a decontamination activity which did not proceed normally. It was determined that a misalignment between SRU prime and SRU backup had occurred when the backup unit was turned on, triggering the FP since the affected AACS design parameter was too sensitive. The fix was to improve the parameter and patch the spacecraft's FSW. This problem could not be uncovered by testing since it could not be modeled in the Cassini test facility.

Safe mode activation #2: During an instrument checkout, Cassini was commanded to perform a slow roll about the Z-axis to keep the X-axis as close as possible to Sun-point while the spacecraft proceeded through Opposition. An overly sensitive AACS control target parameter tripped the Safe Mode response. The SOFS team determined that only flight experience can reveal this problem and the parameter was updated.

Spurious SSPS trip events: Starting at L + 4 months on February 14, 1998, Cassini started to experience trip-off events on its 192 SSPS switches, with an average of two trips per year. Cassini was the first spacecraft ever to use SSPS switches, so that the effect of the flight environment on these devices was not completely predictable. These trips are caused by galactic rays within the flight environment, where one or more photon hits on the voltage comparator of the switch, and can result in a false indication that the current load is anomalously high. This causes the switch to transition from either an "on" or "off" state to a "tripped" condition, which can result in either a benign or serious effect on the spacecraft, depending on which switch trips, and if it is operating at the time of the event.

During the mission, 38 trip events occurred, some of which had significant effects. In May 2005, the USO experienced a trip event, causing loss of communication with the SOFS team until two-way communication could be established once again. In September 2007, the Traveling Wave Tube Amplifier (TWTA) underwent a trip event, causing FP to activate; Safe Mode was executed three times, in addition to a Power-On-Reset (POR) of the RFS system and a Hardware (HW) swap of the TCU and TWTA. The spacecraft's DST was hit in September 2013, causing the Command Demodulation Unit (CDU) to reduce the uplink transmission rate from 500 bits-per-second (bps) to 7.8 bps.

Fix: Nothing can be done to prevent SSPS trip occurrences. Therefore, a new "SSPS Trip" FP algorithm was designed and uplinked into the spacecraft's FSW to address these SEU induced trip events. This new FP monitors each SSPS switch and responds to trips conditions with a predetermined response which is unique for each of the 192 SSPS switches.

Degradation in the ME cover: Shortly before the DSM maneuver, when the ME cover was opened, the cover did not deploy as far as it had in ground tests (14° less than expected), although the opening angle was sufficient to allow for ME burns. The cause was attributed to an increased stiffness in the cover material due to its exposure within the radiation environment of the inner solar system, and to a lesser extent, the long period of disuse. Unfortunately, the ME cover activity within flight environment could not be adequately tested on the ground prelaunch. Since the DSM maneuver, the ME cover opening angle held steady through many cycles, with no further signs of degradation observed. The cover behavior was monitored by the SOFS team until the End of the Mission (EOM).

5.1.3. Outer Cruise

Safe mode activation #3: In 2001, the backup CDS computer experienced a reset due to an oversite in the onboard sequence (human error); a missing telemetry mode definition. As part of the CDS design, all telemetry modes (the rate at which data is downlinked) are executed in both the prime and online CDS computers. As a result of an SOFS exercise to update the SSR with MAG replacement heater patches, one of the backup CDS computer's telemetry modes were overwritten (and thus, was not available), so that it existed only in the prime CDS computer. After activating this particular telemetry mode from the C26 background sequence, the backup CDS reset since the telemetry mode did not exist.

Fix: The SOFS team uses "flight rules" and constraint checklists to ensure errors do not creep into sequences; this particular check was not included in the real-time patch checklist, and was henceforth added to this list.

RWA increased friction anomaly: On December 16, 2000, RWA wheel #2 caused the spacecraft to autonomously switch from RWA control to RCS control due to an increase in friction (triggering FP with no Safe Mode execution), interrupting planned science activities. Analyses determined that this high friction region was localized to the low RPM operating region.

Fix: Constraints were imposed by the SOFS team to avoid the low RPM region for all three RWAs. This was accomplished by altering the wheel speed biasing strategy. A project directive was made to use the RCS system as the primary control for the rest of the cruise phase.

ISS instrument haze anomaly: Five months after the Jupiter flyby event in 2001, it was discovered that a distinct haze was observed around Saturn in images captured by the NAC camera, which had not been seen in previous images. It was determined that this anomaly was caused by contamination of very small particles residing on either the camera's filter assembly or CCD window. It appeared to have been caused by the very long period since the previous decontamination cycle (13 months), and the deeper cold of the environment compared to previous cycles (–90 vs. 0°C).

Fix: A series of decontamination cycles were completed to remove the haze (from periods of 7–57 days in length). A flight rule was added to correct the procedure of heating the ISS camera.

5.1.4. Science Cruise Phase

Safe mode activation #4: The C37 cruise sequence was operating nominally when one of the target vectors was queued to be loaded by the series of commands in operation. Although this target vector was provided in the AACS table being accessed, the associated time-tag associated with the command contained an error, so that it was labeled prior to the start time of the C37 sequence. Since the vector could not be loaded properly, the Safe Mode response was requested.

Fix: The proper vector was reloaded and the sequence restarted onboard the spacecraft; ground procedures were updated to preclude this human error from happening again.

Activation of the redundant RWA #4 wheel: All three RWAs had started to exhibited the same high friction levels at low RPM (drag torque spikes), but unlike RWA #1 and RWA #2, RWA #3 also began to exhibit "cage instability," which is characterized by vibration of the metal cage that holds the ball bearings in place. Analysis showed that the wheel was trending towards possible failure in weeks to possibly months.

Fix: RWA #3 was commanded off to save its remaining life and the redundant (spare) RWA #4 was turned on to replace it.

5.1.5. Saturn Approach Phase

Loss of MAG data during SOI: During the SOI event, no magnetometer data was acquired due to a sequencing error that caused an unexpected instrument reset (instrument FP was triggered). Since SOI was the only opportunity in the prime mission to fly very close to Saturn (until later in the extended mission phases), the loss of science data was considered to be very significant (Figure 8).

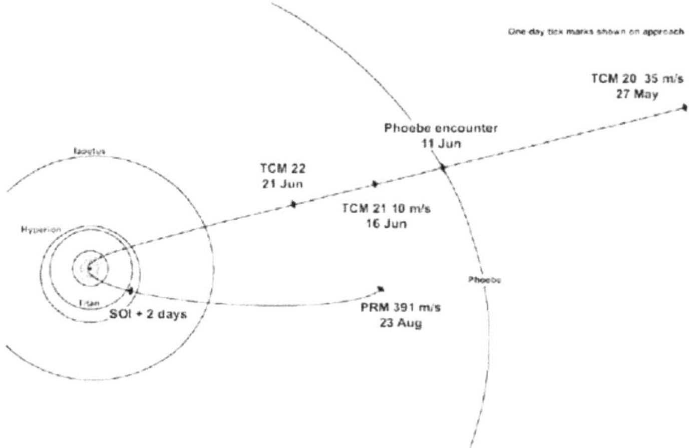

Figure 8. Cassini-Huygens Saturn approach.

5.1.6. Huygens Probe Mission

Probe Doppler bandwidth error: Tests were conducted before reaching Saturn in February 2000 for the Probe ⇒ Cassini ⇒ DSN station data link delivery transmission. These analyses were needed to prepare for the Probe deploy and relay tasks, consisting of several flight exercises and performing "what-if" tests, as well as to validate the Probe's FSW.

Since the Probe's two computers contained minimal onboard data storage capability, the data had to be transmitted to the Cassini orbiter directly during Titan entry, and then relayed to Earth. In this way, the spacecraft would provide the bulk of the data storage needed to support the Probe Relay task, throughout the descent and landing stages of the Probe mission. In the test, the Probe's signal was delivered to the Cassini spacecraft in flight, and then delivered to the DSN station on the ground. Results from this Probe ⇒ Cassini ⇒ DSN station relay test showed insufficient margin to maintain the carrier and subcarrier lock for the duration of the upcoming Probe mission. Analysis showed that the digital circuitry that decodes the data from the subcarrier did not have sufficient bandwidth to properly process the data from the subcarrier once it was Doppler shifted by the expected 5.6 km/s (nominal) velocity difference between Cassini and the Probe. The effect of this anomaly (caused by human error) would yield an unacceptable loss of data during the upcoming Probe Descent ⇒ Titan Landing phase since the digital circuit design did not adequately account for the Probe data's full Doppler shift.

Fix: In January 2001, a joint effort between ESA and NASA established the Huygens Recovery Task Force (HRTF) team to evaluate the problem and develop a solution. This effort leads to a three-part fix that allowed full recovery of the Titan data:

Part 1: The mission profile was redesigned to a Probe trajectory conducive to a low Doppler shift in the Probe-Cassini spacecraft radio link. The early part of the Saturn Tour was redesigned to a higher orbiter flyby altitude of Titan (at 60,000 km). This required that the (original) first two orbital revolutions around Saturn be increased to three revolutions (the tour configuration was unchanged after this point; this extra orbit was at a moderate ΔV cost).

Part 2: Preheating of the Probe's transmitters was necessary before its descent into Titan's atmosphere so that the transmit frequency could be optimized.

Part 3: The new mission design would now have a much lower Doppler shift than that of the original Probe mission. This would require that the Probe be commanded to its "Base Frequency" (referred to as "BITE Mode," a "zero Doppler" test mode that held the lockup frequency at a level equivalent to -1 m/s relative velocity). This BITE Mode of operation must be maintained constantly, even in the presence of fault occurrences and Safe Mode activations. To accomplish this goal, an empty slot within the ATC FSW (eight ATC monitors were in use; four empty placeholders were designed into FSW for future use) was programed to send the "Probe BITE Mode" command continuously, since these ATC algorithms are capable of issuing commands every 12 s, even during and after FP activations.

5.1.7. Tour Operations

Safe mode activation #5: In S33, very soon after a flyby of the Iapetus moon was completed, a SSPS Trip occurred on the prime TWTA. The spacecraft interpreted the SSPS trip as a hardware failure and executed the Safe Mode three times and swapped to the redundant backup TWTA unit. The FP also commanded a TCU swap and an RFS POR.

Fix: The prime TWTA was powered back on (and swapped back), and FSW was updated to implement new FP for selected devices in order to avoid activating SSPS Trip FP. This FSW fix was planned in advance of this incident, based upon observed SSPS trips, but was not uplinked in time to preclude this TWTA SSPS trip.

Exactly 1 year later, another SSPS trip occurred on the prime TWTA unit. FP was not activated due to the new updates.

Loss of MIMI motor drive: In January 2005, a motor controlling one of MIMI's three detectors suffered a mechanical failure. Although all three detectors were still fully functional, one was forced to rely on spacecraft pointing for proper orientation. The loss to MIMI science was approximately 10%.

Loss of SSR DRAM memory: In December 2006, a portion of memory failed within one of the SSR's DRAM memory units, in a location where science/engineering data is stored. This failure was significant because the memory was corrupted, leading to ground software decomposition problems as well as erroneous science and engineering data. No capability to remove or bypass bad areas of SSR hardware memory had been implemented into FSW.

Fix: New capability to bypass corrupted memory locations was uploaded to FSW.

5.2. Extended Mission Experience (Equinox Mission)

RCS thrust branch swap: In March 2009, Cassini swapped over to the backup branch of RCS thrusters to replace those that had been in use since launch, since the prime thrusters were exhibiting increased chamber pressure roughness and decreased thrust (e.g., these thrusters were displaying end-of-life characteristics).

Loss of an ATC temperature sensor: Temperature readings are reported to each ATC from two sensors. For ATC #7, these sensors are mounted on opposite sides of the ME, and are used to monitor chamber temperatures. In 2009, during a maneuver, one of these sensors began to report erroneous data. It was speculated that the failure was most likely caused by a soft short.

Fix: The SOFS team uplinked a command to declare the sensor "dead" (not usable), since the ATC was able to function with only the single remaining sensor. Also, a new operations strategy was developed to eliminate the use of ATC #7 (implemented in S56).

5.3. Second Extended Mission Experience (Solstice Mission)

Safe mode activation #6: On November 2, 2010, during the S64 background sequence, a file was uplinked to reset the backup AFC computer during normal operations. The command was hit by a cosmic ray and corrupted (bit flip), causing the prime CDS computer to reset from receipt of this erroneous command (caused by a failure of the uplinked command to process properly). As a result, Safe Mode was called.

Fix: The chances of a cosmic ray hit on an uplinked command are so unlikely (millions to one), that no fix was implemented.

CAPS instrument failure: In April 2011, Cassini's power bus suffered unexpected swings. The imbalance remained in place until June of that year, when another shift occurred. Engineers suspected the high-rail short to be within the CAPS instrument and 3 days later the instrument was turned off. The bus returned to near preanomaly values, and the CAPS instrument was left off while an investigation was conducted into the cause of the short condition, and whether CAPS could be turned back on. The conclusion was that it was safe to turn the CAPS instrument back on. Two days later the short condition reappeared, causing the bus voltage to shift again. The CAPS instrument was left on and the shifted values remained until June 2012, until a series of voltage swings occurred over a 24-h period. The condition culminated until CAPS was autonomously shut off by the SSPS switch, by an overcurrent draw from the instrument. A second investigation was undertaken after this CAPS anomaly, leading to a decision to leave the CAPS instrument off for the remainder of the mission.

Loss of the USO: At the beginning of the DSN track on December 23, 2011, no downlink signal was received from Cassini. The suspected cause was bad predicts used at the DSN station. New predicts were built and two different DSN antennas were used to acquire the spacecraft's signal to no avail, ruling out the bad predicts as the cause of the anomaly. Attention then turned to the USO as the source of the problem. Cassini's signal was acquired after RTLT (when the USO is no longer used by the spacecraft, but switches over to the DST's VCO).

Fix: A test was devised to determine if the DST's downlink path or the USO was the cause of the loss-of-signal problem. It was determined that the USO had failed. The Auxiliary Oscillator was used for the remainder of the mission, which yielded a "rattier" signal. Spacecraft operations were not affected by the loss of the USO; however, the quality of radio science observations was reduced.

5.4. The Grand Finale

The Cassini mission ended with 20 orbits of the F-Ring, followed by a 22 orbit ballistic trajectory through the D-Ring, and a highly successful final plunge into Saturn. Unique science data was captured during this final flight phase, and no significant anomalies occurred, ending the highly successful, nearly 20-year mission [16].

6. EOM STATISTICS

*Power usage:*Figure 9 shows the entire power history telemetered by the spacecraft during its 20-year mission, including the very last data point sent just before EOM. The overall RTG power decay shows an exponential behavior starting from 882.1 W on the Day 1, to 600.3 W on the last day of the mission (indicating a total power decay of 32%). The data plot indicates nominal RTG performance, with some peculiarities: during the first 3 years of the mission, the power output decayed by 70 W at an accelerated rate due to the dopant precipitation in the SiGe thermocouple, which reduces the available current carriers.

Cassini's SSRs used as a radiation detector: As discussed above, Cassini's SSRs were susceptible to high SBE and DBE occurrences due to environmental effects. These elevated bit error counts often occurred in the presence of high dust and radiation. In this way, the SSRs were inadvertently turned into uncalibrated and unofficial radiation detectors. Figure 10 shows the effects of radiation on the SSRs during several ring flybys, relative to the SBE error count. The spacecraft's flybys between the D-Ring and Saturn's atmosphere are shown as the "proximal orbit" region. These results were consistent with MIMI's radiation model.

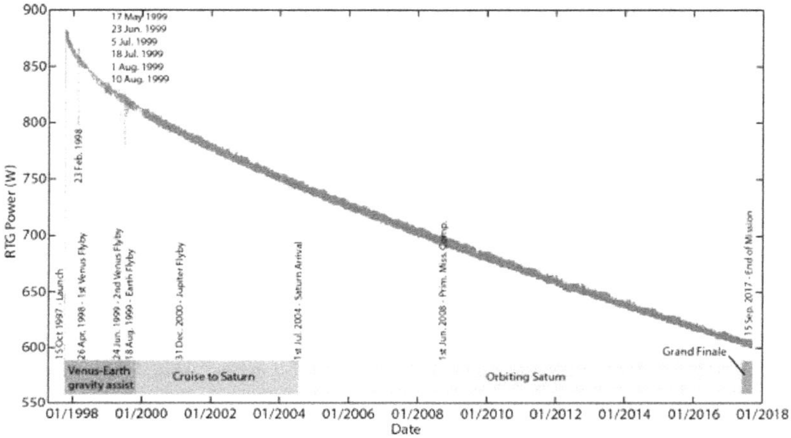

Figure 9. Cassini recorded power usage over its 20 year mission (credit: Grandidier et al. [17]).

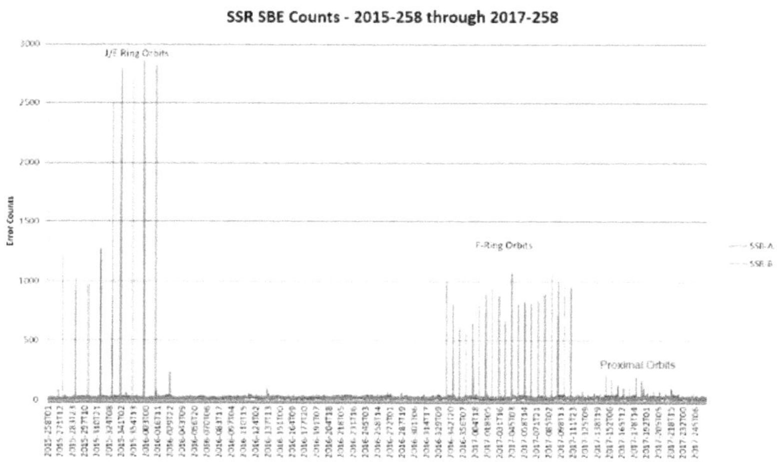

Figure 10. SSR SBE counts from late 2015 through EOM (courtesy of S. Adamiak).

7. SATURN SCIENCE

Many incredible discoveries were uncovered by the Cassini-Huygens mission during its 20 years of flight; a few of those fascinating encounters are mentioned here. Figure 11 depicts six of the more than 60 of Saturn's known moons, which range in size from a few hundred meters to larger than planet Mercury. The top row of this figure, from the left (not to scale), shows the tiny odd looking moon Pan, Mimas (which looks like the "Death Star" space station from movie Star Wars), and Hyperion, which resembles a sponge. On the bottom row are Iapetus, Titan (the largest of Saturn's moons), and Enceladus, which contains "tiger stripe" fissures with erupting plumes, implying an underground reservoir of water that is suspected to be around 10 km deep (i.e., an underground ocean). Figure 12 depicts an artist's impression of the hydrothermal activity taking place on this south polar region of Enceladus. Hot water traveling upward from the ocean comes into contact with cooler water, which is eventually expelled through the vents that connect the ocean to the surface of the moon.

Figure 11.Six of Saturn's unique moons (image credit: NASA/JPL-Caltech/Space Science Institute).

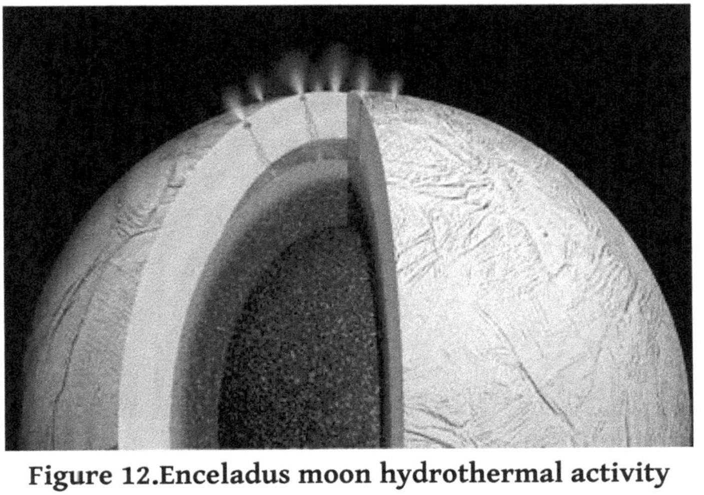

Figure 12.Enceladus moon hydrothermal activity

(image credit: NASA/JPL-Caltech).

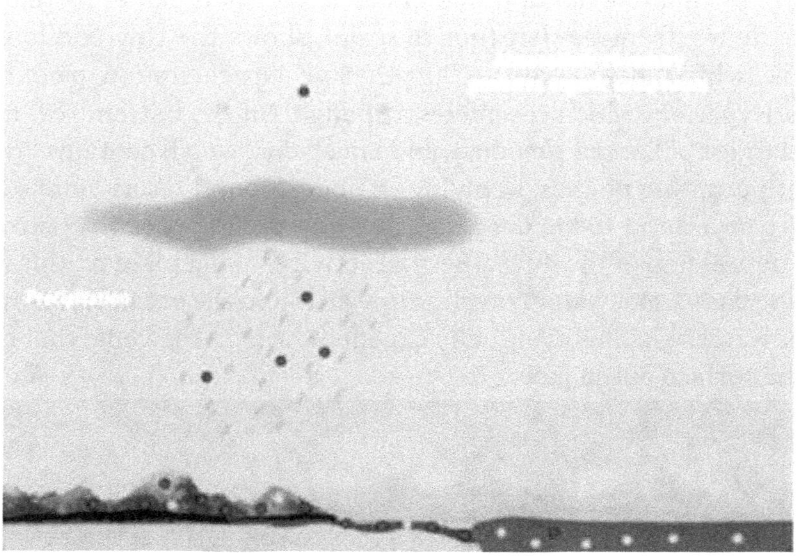

Figure 13. Titan's sea and lake organic compounds.

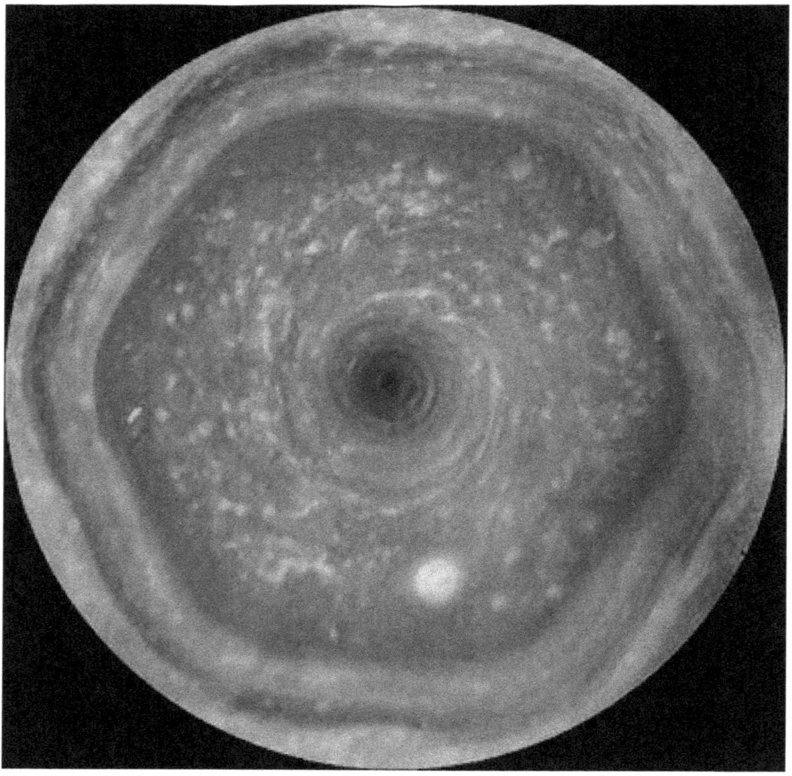

Figure 14. Saturn's polar hurricane (NASA/JPL-Caltech/SSI/Hampton University).

Figure 13 depicts how different organic compounds make their way to Titan's seas and lakes. Ligeia Mare is one of three of Titan's seas, consisting of pure methane and a seabed covered by sludge-like organic-rich material. Titan's atmosphere of nitrogen and methane react to produce organic molecules, the heaviest of which fall to the surface through air and rainfall, some of which make their way to the sludge on the sea floor. Figure 14 depicts the giant hexagonal hurricane at Saturn's north pole (approximately 30,000 km across). An intense six-sided jet stream with winds at 320 km/h spirals around a massive storm which rotates anticlockwise at the heart of this region (false color image).

8. CONCISE SUMMARY

For robotic spacecraft to complete their goals successfully without significant risk or degradation to mission objectives, preventative measures for instruments and subsystems must be implemented by way of a robust FP strategy and onboard FSW flexibility. Prelaunch analyses and tests conducted to preclude problems do not always safeguard against human error, the flight environment, or design oversights, nor can they capture all fault cases. Mission planners must acknowledge that unknown problems can still surface after launch. During the Cassini-Huygens mission, this was proven true by the need for several new FP routines, FSW updates, and FSW patches required to resolved unexpected problems not anticipated by prelaunch designers. For interplanetary spacecraft like Cassini, these fixes were made more manageable given that significant time was available during the cruise phase to augment the FP and patch FSW in order to address these unforeseen problems, due to the flexibility that designers built into the FSW architecture.

9. AT A GLANCE

Robotic interplanetary spacecraft sent to the outer planets of our solar system encounter various obstacles, including managing spacecraft subsystems, coping with material stresses from solar heating near Earth and the cold of deep space, dealing with solar radiation and cosmic rays, ensuring sufficient power for engineering devices and scientific instruments, addressing time-sensitive onboard issues due to long round-trip light time, and safeguarding critical activities like moon/planet flybys, probe deployment, and scientific objectives. This chapter describes the strategy used on the Cassini Mission-to-Saturn spacecraft, including how its onboard subsystems are protected and maintained, the benefits of automated fault protection routines, protocols to prevent human errors in uplinked sequences, and updating flight software based on new discoveries about the flight environment. These measures ensure that mission objectives are achieved despite the increasing delay between ground commands and the spacecraft as it nears Saturn, while also safeguarding planetary protection constraints during the spacecraft's final descent into the planet.

REFERENCES

1. Doody D. Basics of Space Flight. 2004. Available from: http://www2.jpl.nasa.gov/basics/ [Accessed: 2018-08-24]
2. Saturn Tour Highlights Cassini-Huygens [Internet]. 2018. Available from: https://saturn.jpl.nasa.gov/ [Accessed: 2018-08-24]
3. Taylor J, Sakamoto L, Wong C. Cassini Orbiter/Huygens Probe Telecommunications Descanso Design and Performance Summary Series [Internet]. 2002. Available from: https://descanso.jpl.nasa.gov/DPSummary/Descanso3--Cassini2.pdf [Accessed: 2018-08-24]
4. NASA Office of Planetary Protection [Internet]. 2014. Available from: http://planetaryprotection.nasa.gov/ [Accessed: 2018-08-24]
5. Morgan P. Cassini spacecraft-DSN communications, handling anomalous link conditions, and complete loss-of-spacecraft signal. In: Dekoulis G, editor. Space Flight. 1st ed. Croatia: InTech-Open Access; 2018. pp. 223-250. ch12
6. Morgan P. Robotic spacecraft health management. In: Johnson S, Gormley T, Kessler S, Mott C, Patterson-Hine A, Reichard K,et al., editors. System Health Management: With Aerospace Applications. 1st ed. New York, NY: Wiley; 2011. pp. 543-554. ch34
7. Smith K, Peng C. Modal test of the Cassini spacecraft. In: Proceedings of SPIE; Issue 3089/Part 1. Orlando, Florida: The International Society for Optical Engineering; 1997. pp. 804-810
8. Scharton T, Chang K. Force limited vibration testing of the Cassini spacecraft and instruments. In: 17th Aerospace Testing Seminar; Manhattan Beach, California; Volume 17. October 1997. pp. 85-94
9. Coleman M, Peng C, Smith K. Test verification of the Cassini spacecraft dynamic model. In: IEEE Aerospace Conference Proceedings. Vol. 1. 1997. pp. 289-302
10. Himelblau H, Hughes W, McNelis A, Kern D, Bergen T. Derivation of acoustic criteria for the Cassini spacecraft and comparison with flight data. In: 6th International Congress Proceedings (Sound and Vibration); Lyngby, Denmark; Volume 7. 1999. pp. 3167-3174
11. Mireles V, Tsuyuki G. A summary of the Cassini system-level thermal balance test: Engineering subsystems. SAE Transactions. 1997;106:940-953. American Technical Publishers LTD

12. Tsuyuki G, Mireles V, Lin E, Avila A. A summary of the Cassini system-level thermal balance test: Science instruments. In: Proceedings of the 27th Environmental Systems International Conference; Lake Tahoe, Nevada. 1997

13. Morgan P. Enhancing the Cassini mission through FP applications after launch. In: Proceedings of the IEEE/AIAA Conference; Big Sky, Montana; March. 2016

14. Swift G, Guertin S. In-flight observations of multiple-bit upset in DRAMs. In: Proceedings of the 37th Nuclear and Space Radiation Effects Conference, vol. 47(6); Reno, Nevada. 2000. pp. 2386-2391

15. Morgan P. Cassini Spacecraft's In-flight fault protection redesign for unexpected regulator malfunction. In: IEEE/IAAA Conference; Big Sky, Montana; March. 2010

16. The Grand Finale Toolkit [Internet]. 2018. Available from: https://saturn.jpl.nasa.gov/mission/grand-finale/overview/ [Accessed: 2018-08-24]

17. Grandidier J, Woerner DF, Burk TA. Cassini power during the 20-year mission and until the final plunge into Saturn. Journal of Spacecraft and Rockets. 2018. DOI: 10.2514/1.A34291

7 Chapter ADVANCED NONLINEAR MODELING OF GAS TURBINE DYNAMICS

1. INTRODUCTION: AN OVERVIEW

The gas turbine (GT) is a powerful source of energy that has relatively low size and weight. It is a principal power plant for aviation and electric energy production and has many other successful applications. For instance, aero-derivative engines are widely used for electricity generation in offshore platforms and as marine power plants because these engines are more compact and have faster dynamics than industrial GTs. A significant growth of a GT industry has been observed in the last decades [1].

1.1. Gas Turbine Modeling

As new gas turbines are being developed, there is an increasing use of mathematical modeling and simulation in designing these engines and their systems. Developing gas turbine models has shown to be a successful approach in the design and manufacturing process. Aside from engine development, GT modeling and simulation have various other uses, including control system design, condition monitoring, problem detection, and system identification. For instance, it allows for recreating the performance of a certain engine by fitting a model to experimental data gathered in test beds or outdoor circumstances. By utilizing a more precise individual engine model rather of a generic one, control and diagnostic algorithms can be enhanced [2].

Technical system models, such as GT models, can be categorized as linear or nonlinear and as data-driven (black-box models) or physics-based (white-box models). In spite of wide application of simplified linear modeling and simulation of GTs, the behavior of these machines is usually nonlinear, and precise nonlinear models are unavoidable [3].

Data-driven models do not require intricate system understanding to create accurate models. They rely on existing empirical data and are guided by optimization techniques or, in the case of artificial neural networks, by machine learning. Due to their simplicity, these models are commonly utilized in gas turbine design. A comprehensive overview of various data-driven models for gas turbines can be found in a book. [3].

Physics-based modeling utilizes the physical rules governing turbo-machines to accurately simulate their behavior. These models are intricate and less commonly utilized. Nevertheless, they hold complex information that is challenging to extract from empirical data and are commonly used as a foundation for developing more straightforward data-driven models. Physics-based modeling is a primary form of mathematical modeling for gas turbines.

1.2. Thermodynamic Model

The above reasoning explains why a component-based nonlinear gas turbine model is considered as principle for the design of the engine itself and for developing its control and monitoring systems. It is a highly complex thermodynamic model based on the aerothermal calculations of a gas path and the description of engine's components (compressor, combustion chamber, turbine, etc.) by nonlinear performance maps. Foundations of the thermodynamic models can be found in [2, 4].

The thermodynamic model comprises two interrelated parts, namely nonlinear static model and nonlinear dynamic model (NDM). The nonlinear static model allows investigating steady-state performances of the engine before its final creation. This model may include more than 100 algebraic and transcendent mathematical relations and, in general, presents a system of nonlinear equations (see [2]). These equations reflect the mass and energy balance between engine components during stationary operation, and the number of such equations typically varies from 5 to 15.

Once the static model has been created, the detailed nonlinear dynamic model can be developed with fewer efforts because it is similar to the static model. The principal difference is that the mentioned algebraic equations of mass and energy balance at steady states are now written in the form of differential equations at transients. The number of such equations corresponds to the number of mass and energy accumulators simulated. Since NDM is a complex and relatively slow procedure, many simplified models are constructed on its basis to be used for the aims of engine control and diagnosis. Nevertheless, along with increasing processor and developing the methods of execution time minimization, direct use of the NDM in real time is becoming possible [2].

The development and use of the thermodynamic models have started in the 1970s, in many respects, by the studies of Saravanamuttoo et al. (for example, [5]). Since that time, many improvements related to higher accuracy and more detailed engine's component description were introduced in this model; some of them are mentioned below.

Stamatis et al. proposed in [6] the scheme of adaptive simulation by a nonlinear system identification technique that later was used for multipoint gas turbine diagnosis [7]. Ellipsoid functions were introduced in [8] for more accurate description of the components' maps and better identification of a whole engine at steady states and transients. The authors of paper [9] developed a stage-based compressor model to be used in the thermodynamic model instead of a compressor performance map. This modified thermodynamic model allows the localization of the faulty stages of a multistage compressor and identification of the three compressor degradation mechanisms: fouling, tip clearance increase, and erosion of aerofoils. Thus, gas turbine diagnostics become more profound. Since the early 1990s, Joachim Kurzke has developed the universal program GasTurb for nonlinear physics-based gas turbine simulation [10, 11]. This commercial software allows simulating different types of engines and helps to solve various design and analysis problems. The program GasTurb has special tools to analyze, correct, and enhance the component maps contributing in this way to the accuracy of final engine simulation. Another way to improve the simulation accuracy is proposed by Volponi et al. [12]. As an engine measurement system has individual systematic measurement errors, the authors propose to compensate them by an additional data-driven model on the basis of artificial neural networks. The introduced hybrid model is constructed from a traditional thermodynamic model and this data-driven model. It is shown that the hybrid model can more accurately simulate the performance of a particular engine than the thermodynamic model itself.

The above improvements are related to a static part of the thermodynamic model or both static and dynamic parts. However, the description of engine transients has specific problems to solve, and their solution can additionally improve the dynamic part, namely, detailed nonlinear dynamic model. For gas turbine control and monitoring systems as a whole and, more importantly, for the systems of aircraft gas turbine engines, accurate and fast NDMs are in high demand [2, 13]. These detailed nonlinear models will be useful for the implementation of model predictive control and more effective diagnosis at transients where simplified Kalman filter-based techniques have often been used to date [14]. Since here, this chapter will deal only with such models.

Modern NDMs generally take into considerations three "accumulators":

- mass and energy accumulation in pneumatic gas path volumes,
- mechanical energy accumulation in the rotors,
- heat accumulation in the stator and rotor heated parts (disks, blades, vanes, case elements).

The volume dynamics is very fast and it is important for controller design. The rotor dynamics lasts for aircraft engines about 10–15 s and has the largest influence on engine performance. The heat exchange dynamics may last many minutes but its direct influence on gas path variables is small because the heat interchange between gas flow and engine-heated parts (HPs) is by far smaller than total energy of the gas. These reasons explain why the models that simulate the rotor dynamics only are still used in diagnostics.

1.3. Tip Clearance Dynamic Effect

There is also an important indirect dynamic effect of the warming-up of engine parts, and this effect has not been taken into consideration in the NDMs yet. The point is that during engine transients, the dynamics of the warming-up is different for a rotor and a stator. The rotor parts, especially massive disks, change slowly their temperature while relatively thin stator parts are warmed up faster. Consequently, the radial displacements of rotor blade tips delay from those of the corresponding casing surfaces, and tip clearances dynamically increase during engine acceleration and decrease during the deceleration.

As the result of increased tip clearance losses, the efficiency of compressor and turbine components lowers and overall engine performance significantly degrades. Sobey and Suggs 1963 demonstrate in book [15] that the 1% turbine tip clearance increase results in the 1% reduction of turbine efficiency and the 1.5–2% increase of engine-specific fuel consumption. The impact of a compressor clearance is even greater: the 1% clearance increase causes the 1.5–3% specific fuel consumption growth. As shown in [16] for the acceleration from the idle to the take-off regime, the increase rate of aircraft engine thrust can reduce twice due to the dynamic clearance increase. The maintenance results show that the corresponding thrust loss can reach from 3 to 15% and takes place from 20th to 60th second after the engine regime change. Thus, significant thrust reduction can happen during the aircraft take-off putting the flight at risk.

Since modern aircraft gas turbine engines need effective control and monitoring systems, accurate detailed nonlinear dynamic models are in increasing demand. In this way, the modeling of the above-described dynamic clearance effect must be implemented in NDMs. So far, such dynamic models use fixed component performance maps obtained at steady states for warmed-up components. The difficulty to introduce the effect of dynamic clearances consists in the fact that they depend on stress-strain state of the stator and rotor parts, and the stresses and strains have irregular distribution that varies in time. Thus, it becomes clear that accurate modeling of the dynamic clearance effect needs the application of finite element methods to the heated parts. The problem is that such calculations are very time-consuming and cannot be directly implemented in NDMs.

As mentioned above, the clearance effect can be modeled only by the analysis of stress-strain state of both stator and rotor HPs with a known initial temperature distribution and a heating rate. The clearance model that meets such conditions was considered in [17]. Unfortunately, this model does not take into account a real shape of HPs and therefore cannot ensure a high accuracy of dynamic engine simulation. The model presented by Archipov et al. in [18] already accounts for the shape but takes other strong limitation that the material properties are independent on material. The authors also make a disputable statement that aerodynamic and pressure gas forces have significant influence on the clearances only for high-power low-speed turbines of industrial power plants. Kurzke proposed NDM with the dynamic clearance model integrated [19]. However, this model does not take into account a radial disk extension and the aerodynamic action of gases on the blade and the casing. Paper [20] compares three variations of the dynamic clearance model. It was found that impulse response model is the most accurate but also time- and memory-consuming. Thus, this model cannot be directly integrated into NDM.

Our previous papers [21, 22] generally follow the ideas of the impulse response model. First, solid models of hot parts of a turbofan had been created. Second, multiple calculations were conducted using the finite element method to understand how the displacements of the hot part surfaces depend on the external temperatures and the loads applied. Third, on the basis of these numerical experiments, a simplified dynamic clearance model (SDCM) was formed. Forth, SDCM was integrated into a nonlinear dynamic model of the turbofan resulting in an enhanced nonlinear dynamic model (ENDM). The rest of this chapter is devoted to the results of the mentioned finite element method calculations, description of the SDCM and its integration into NDM, and the results of simulation by the enhanced model. A high-pressure turbine and its disk are mostly used to exemplify the proposed methodology.

2. ENHANCED NONLINEAR DYNAMIC MODEL

The structure and operation of the above-mentioned ENDM are illustrated by Figure 1. Modules 1–4 constitute SDCM of an engine component, compressor or turbine. Module 5 presents an initial engine model, which, in conjunction with SDCM, presents the enhanced nonlinear model. Only one component is presented on this scheme for simplicity. The software of ENDM includes SDCMs for all the components where the dynamic clearance effect is significant. The enhanced model has been developed for a low-bypass two-spool turbofan engine of a maneuverable aircraft. All simplified relations for the clearance model were obtained through exhaustive calculations in ANSYS with the solid models of engine stator and rotor hot parts, namely disk, blade, and casing. The below description of the clearance model is given for a high-pressure turbine (HPT) of this engine as a component example.

The initial nonlinear dynamic model includes the dynamics of two engine rotors. The corresponding differential equations are solved through their integration by an iterative procedure. In each step, the corrections to state variables (rotation speeds) are obtained and all engine variables are renewed.

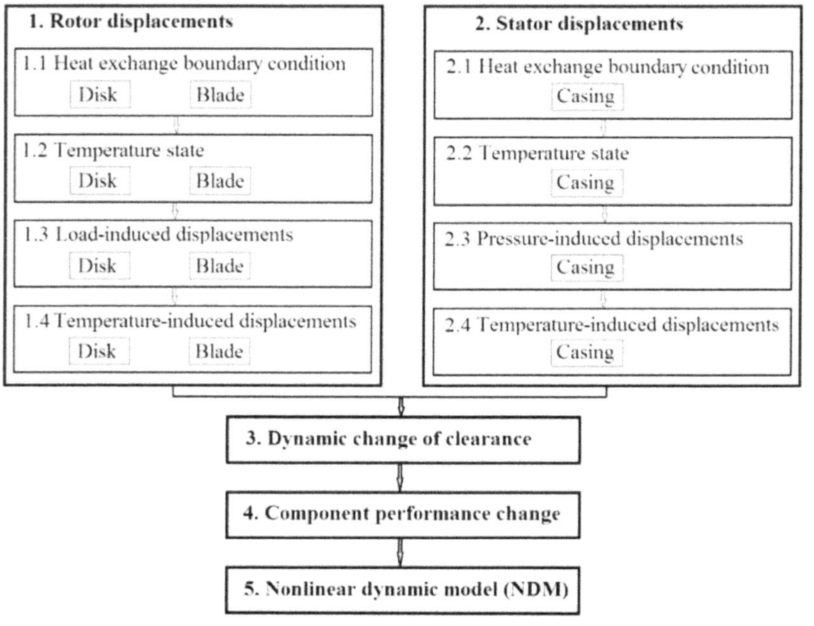

Figure 1. Structure of the enhanced nonlinear dynamic model.

The enhanced engine model conserves this iterative character. At an actual step, the variables computed by NDM are used in SDCM to calculate a new tip clearance and corrections to component performances. The modified performances are employed at the next step. In this way, as with real engine dynamics, ENDM uses component performances that are dynamically changed.

Blocks 1.1–1.4 for the stator and Blocks 2.1–2.4 for the rotor illustrate what happens with heated parts (disk, blade, and casing) and tip clearances when engine operating mode dynamically changes. After the mode change, the gas path variables calculated by NMM (Block 5) begin to vary producing the change of heat exchange boundary conditions outside of the heated parts (Blocks 1.1 and 2.1). Because of heat accumulation or loss, the temperature state (distribution of metal temperature) of HPs begins to change (Blocks 1.2 and 2.2). Elevated temperatures of HP cause its thermal expansion. As the HP temperature state has a delay relative to the engine mode change, the corresponding displacement (Blocks 1.3 and 2.3) varies with a delay as well. The knowledge of the temperature state also allows us to correctly consider the action of forces on the HP radial displacements. As shown in [21], for a disk and blade, significant displacements are caused by a centrifugal force, while a pressure force is the most influencing for a casing. These forces are considered in Blocks 1.4 and 2.4 accordingly. When an engine operating point is changing, the force applied to HP changes as well, and the force-induced displacement reacts immediately. However, the engine mode variation also means the change of the heat exchange boundary conditions resulting in other temperature state, other metal elasticity, and an additional change of the displacement. Thus, the force-induced displacement has a static component that immediately reacts on the engine mode and a dynamic component that reacts with a delay. In this way, the total radial displacements of the surfaces that form a tip clearance have a complex dynamic behavior.

The calculation of a dynamic clearance change (see Module 3), which is necessary to evaluate the change in a component performance, is based on simple relations. An actual dynamic tip clearance δ that is a function of transient time can be expressed through a clearance δ0 of a cold turbine and total displacements uC,uD,δB of the casing, disk, and blade accordingly, resulting in

$$\delta = \delta_0 + u_C - (u_D + u_B) \quad (1)$$

Since the HPT performance map used in NDM corresponds to the turbine parts completely warmed up at steady states, the necessary dynamic correction of the performance will depend on a difference Δδ between a dynamic clearance δ and a static clearance δst. Let us express the static clearance in the form of Eq. (1) but using static displacements of HPs. This yields:

$$\delta^{st} = \delta_0 + u_C^{st} - (u_D^{st} + u_B^{st}) \quad (2)$$

In this way, the dynamic clearance change is written by:

$$\Delta\delta = (u_C - u_C^{st}) - (u_D - u_D^{st}) - (u_B - u_B^{st}) \quad (3)$$

Using the clearance change Δδ as an input parameter, Module 4 corrects the component efficiency η because it is known that just this performance is affected by an increased clearance. Paper [23] shows that a turbine efficiency loss Δη is linearly dependent on a relative clearance change:

$$\Delta\bar{\delta} = {}^{\Delta\delta}/_{L} \quad (4)$$

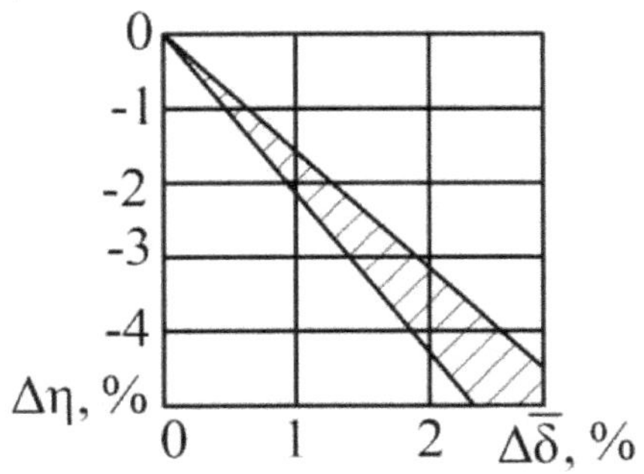

Figure 2. Turbine efficiency losses vs. an increasing tip clearance.

where L is a blade length. Figure 2 illustrates this relation.

Therefore, Module 4 computes the corrected component efficiency at each point of a transient process according to an expression:

$$\eta = \eta^{st} - k\Delta\overline{\delta} \qquad (5)$$

where ηst is the efficiency at an equivalent steady state and k is the coefficient depending on the construction of a modeled engine and its component.

Since Module 5 employs the corrected efficiencies of all the engine components as input parameters, the engine variables simulated by NDM take into consideration the effect of dynamically varying radial clearances.

As can be seen from the above description, the equations of Modules 3 and 4 are simple and do not need additional explanations. A nonlinear dynamic model of Module 5 is not simple, but it does not need additional description because this type of gas turbine models is well described in literature. However, Modules 1 and 2 and their blocks were presented in this section in a schematic form necessary for a general understanding of the enhanced NDM. As described before, in these modules, the displacements of the heated parts are calculated. To know how these displacements depend on external factors, the displacements were simulated in ANSYS by creating the solid model of each HP and by applying the finite element method to determine the HP stress-stain state. By multiple numerical experiments of this type, the simplified relations between the displacement and external dynamic factors were formed and included in Modules 1 and 2. This ANSYS-based simulation of the HP displacements is described in the next section.

3. FINITE ELEMENT-BASED DISPLACEMENT SIMULATION

3.1. Thermal Boundary Conditions and Mechanical Loads

As shown in Figure 1 and described in Section 2, the displacements of the heated parts depend on thermal boundary conditions and mechanical loads applied. Shown in Figure 3, the design scheme of the disk, which is the most complex HP, illustrates these conditions and loads.

The boundary conditions are set by an external air and gas temperature T and heat transfer coefficient at the disk surface. Since these conditions considerably vary at different parts of the surface, it is broken down into 24 sections with constant temperatures Ti and coefficients i.

At one engine steady state called a reference mode, the values T0i and 0i of these parameters are known on the basis of the experimental information. In a peripheral disk part, in addition to hot gases, heat is transmitted from the blades. This additional heat transfer is taken into account by elevated values T1 and 1 in Section 1 of the disk surface.

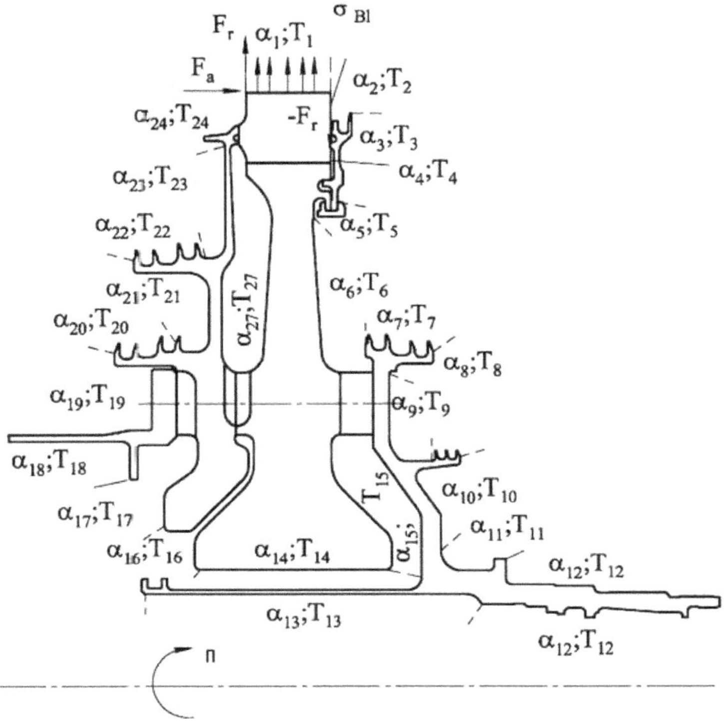

Figure 3. Design scheme of disk thermal boundary conditions and mechanical loads.

As to the mechanical loads, the centrifugal force acting on the disk is a body force that is applied to each elemental volume of the disk. The centrifugal force from the rotating blades is given as a surface force by a uniform distribution σB in Section 1. The design schemes of the blade and the casing are similar.

3.2. Stress-Strain State and the Displacements of Heated Parts

In the finite element-based simulation, the heated parts are presented by their solid models illustrated by Figure 4. In its solid model, each HP is divided on elemental 3D simplex volumes. Each volume is presented in finite element calculations by four nodes.

After the determination of the temperature state of a heated part, the nonuniform distribution of its temperature t is known. In addition to the action of this temperature, the heated part undergoes the action of a surface force p(x,y,z) and a body force F(x,y,z) . The known temperature irregularity and the forces induce in each node a displacement u=[uxuyuz]т , a strain ε=[εxεyεzγxyγyzγzx] т, and a stress σ=[σxσyσzσxyσyzσzx]т that are described by the following linear equations of the elasticity theory (see [24]):

$$\varepsilon = \mathbf{Ru}; \qquad (6)$$

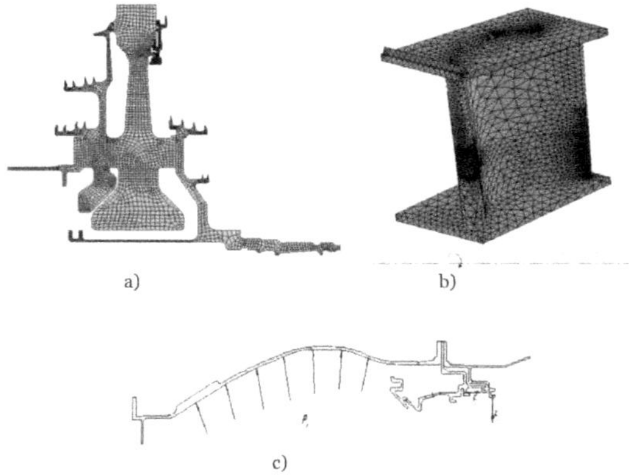

Figure 4. Solid models of the heated parts of HPT (a—disk, b—blade, c—casing).

$$\sigma = D(\varepsilon - \alpha t); \quad (7)$$

$$R^T\sigma + F = 0 \quad (8)$$

and by the equation of boundary condition:

$$p - C\sigma = 0 \quad (9)$$

In these equations, **R** is a differential matrix operator, **D** presents a stiffness matrix depending on material elasticity and the Poisson's ratio, denotes a linear expansion coefficient vector, and **C** stands for a rotation matrix. On the basis of Eqs. (6)–(9) of one elemental volume, a huge system of linear equations of a whole heated part is formed. The number of unknown variables in this system can be partly reduced because one volume node pertains to some adjacent elemental volumes. The system is solved by the least squares method. As a result, the displacements of the external surface of the heated element are determined separately for the action of thermal expansion and the force. In addition to the reference engine mode, the displacements of all the heated parts were determined at the idle regime.

As mentioned in Section 3.1, the thermal boundary conditions are known at the reference mode and the variables necessary to determine mechanical loads at this mode are simply calculated by NDM. Using these data, the displacements u0t and u0F induced at this mode by temperature and force were firstly computed in ANSYS for the disk and the other heated parts. To know how these displacements vary during engine operation, let us firstly analyze how the thermal boundary conditions depend on an engine operating mode. It will be shown that the boundary conditions can be determined through actual and reference gas path variables known from NDM.

4. VARYING BOUNDARY CONDITIONS

As mentioned above, the values of the boundary parameters Ti and i in the sections of the HP surface (see Figure 3 for the case of the disk) are known only for the reference mode. To have the possibility to make the finite element calculation in ANSYS at any mode, we need to know how these parameters vary along with an engine operating point.

4.1. Boundary Temperatures

Oleynik has shown in his thesis [25] that the distribution of boundary temperatures around HP at a current operating mode is similar to the distribution at a reference mode. The calculations made with NDM of the engine under analysis also confirm that gas path temperatures proportionally change from one operating point to another [22]. In this way, we can state that a temperature similarity coefficient $k_T = (T_i - T_{HPC}) / (T_i^0 - T_{HPC}^0)$ is approximately constant and a current temperature at any section "i" of the HP surface can be expressed through this coefficient by:

$$T_i = k_T \cdot (T_i^0 - T_{HPC}^0) + T_{HPC} \quad (10)$$

The similarity coefficient is determined using the gas path temperatures computed by NDM at the reference and actual engine modes.

4.2. Heat Transfer Coefficients

Paper [22] shows that the heat transfer coefficients i change proportionally when an operating mode varies. Using known relations between different criteria of gas flow, this chapter derives the following equation for a similarity coefficient:

$$k_\alpha = \frac{\alpha}{\alpha^0} = \left(\frac{n}{n^0}\frac{P_{HPC}}{P_{HPC}^0}\right)^{0.8}\left(\frac{T_{HPC}}{T_{HPC}^0}\right)^{-0.567} \quad (11)$$

As the necessary actual and reference values of gas path variables are known from NDM, the similarity coefficient is simply calculated and the coefficients i at the HP surface sections are determined by:

$$\alpha_i = k_\alpha \alpha_i^0 \quad (12)$$

In this way, the distribution of the boundary variables T and i=k 0i can be simply determined through the NDM gas path variables, namely, HPT rotation speed n, high pressure compressor (HPC) discharge temperature THPC , and HPC discharge pressures PHPC .

The next challenging problem was to create the relations for calculating the HP displacements, both temperature induced and force induced, at any engine dynamic operating point. Let us begin from the displacements due to thermal expansion of the heated parts.

5. VARYING THERMAL EXPANSION DISPLACEMENTS

The dynamics of the displacements caused by temperature loading is described below using a displacement transient performance. To determine this performance, the influence of a step change of boundary temperatures from 293 K (cold disk) to the distribution at the reference mode was simulated in ANSYS. It was found that, in addition to time τ, the displacement also depends on the heat transfer similarity coefficient k , and the displacement performance was presented as a relative function $\bar{u}(\tau, k_\alpha) = \frac{u(\tau, k_\alpha) - u^0}{u^{st} - u^0}$ illustrated by Figure 5.

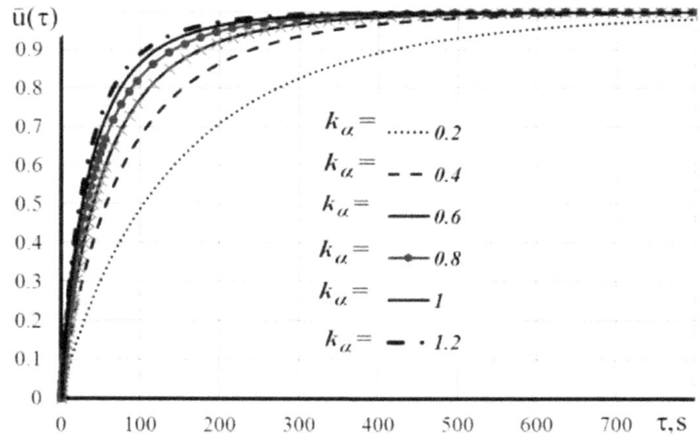

Figure 5. Transient performance of a disk displacement.

Figure 6 shows the transient performances of the blade and casing absolute displacements obtained in ANSYS by the same mode. For these heated parts, the influence of the coefficient k is insignificant.

Using the disk as an example, let us now show how to consider its displacement performance in a total process of the ENDM computing. Paper [22] demonstrates that, for each value k , the corresponding curve in Figure 5 is accurately described by a weighted sum of two exponents and therefore can be presented by:

$$\bar{u}(\tau, k_\alpha) = \sum_{j=1}^{2} k_j(k_\alpha)\left(1 - e^{\frac{\tau}{T_j(k_\alpha)}}\right), j = 1, 2 \qquad (13)$$

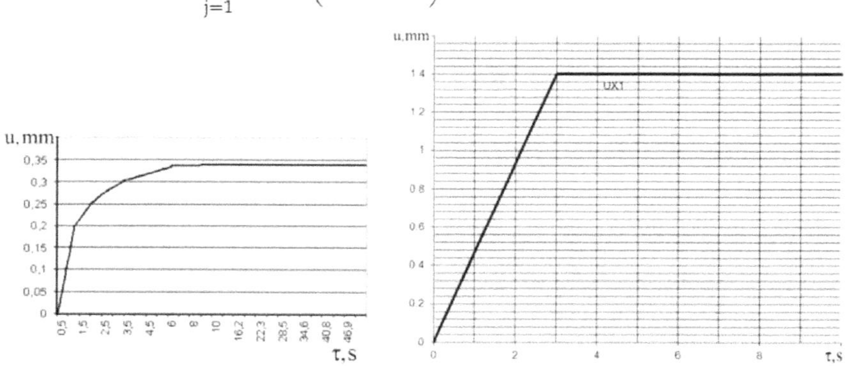

(a) (b)

Figure 6. Transient performance of blade and casing displacements (a—blade, b—casing).

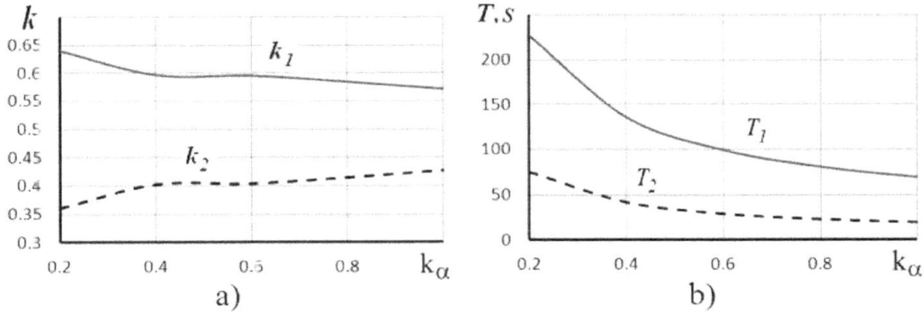

Figure 7. Weighting coefficients and time constants vs. heat transfer similarity coefficient [21] (a—weighting coefficients, b—time constants).

where kj is a weighting coefficient and Tj is a time constant. For each value of k , four parameters k1, k2, T1, and T2 were determined. Figure 7 illustrates their dependency on the coefficient k .

The two mentioned exponents present analytical solutions of linear differential equations that for absolute displacements take a form:

$$T_j(k_\alpha)\frac{du_j}{d\tau} + u_j = u^0(k_\alpha),\ j = 1, 2 \quad (14)$$

The following two equations:

$$u_j(\tau, k_\alpha) = \frac{T_j(k_\alpha)}{T_j(k_\alpha) + \Delta\tau} u_j(\tau - \Delta\tau) + \frac{\Delta\tau}{T_j(k_\alpha) + \Delta\tau} u^0(k_\alpha),\ j = 1, 2 \quad (15)$$

are numerical solutions of these equations. Their weighted sum:

$$u(\tau, k_\alpha) = \sum_{j=1}^{2} k_j(k_\alpha) u_j(\tau, k_\alpha),\ j = 1, 2 \quad (16)$$

is a final expression to numerically compute the dynamic displacement caused by thermal disk expansion. Eqs. (15) and (16) present final steps in Block 1.3 of the enhanced nonlinear dynamic model (see Figure 1). Through the coefficient k , the displacement calculation is adapted to an actual dynamic engine operating point. The blade and casing displacement (Blocks 1.3 and 2.3) are computed similarly.

6. VARYING FORCE-INDUCED DISPLACEMENTS

The displacements induced in HPs by mechanical loads can be considered elastic and proportional to the load. For the disk and the blade, the main load is a centrifugal force and the displacements will be proportional to the rotation speed squared n2. As the casing is mainly loaded by a pressure force, the displacement will linearly depend on the HPC pressure PHPC . The action of these forces has no delay and the displacement will change along with the load change.

However, since the elasticity coefficient depends on the HP temperature, the HP displacement should be simulated regarding this dependency. The temperature distribution within HP is nonuniform and dynamically changes during transient engine operation. For this reason, it will be difficult to directly simulate the elasticity change. To solve this problem, paper [22] proposes the concept of an equivalent temperature.

6.1. Equivalent Temperature

The equivalent temperature te is defined as a temperature of a uniformly heated engine part, which has load-induced displacements equal to the displacements of HP with an actual temperature state and the same mechanical loading. Using the temperature te, the displacement at an actual dynamic point is written for the disk and blade by:

$$u_F = u_F^{\circ}(t_e) \cdot \left(\frac{n}{n^0}\right)^2 \quad (17)$$

and for the casing by:

$$u_F = u_F^{\circ}(t_e) \cdot \left(\frac{P_{HPC}}{P_{HPC}^0}\right) \quad (18)$$

The displacement u∘F corresponds to a hypothetical situation when HP is under the constant mechanical load of the reference mode, but the HP heating conditions are varying and correspond to the actual engine operating point. A function u∘F(te) was determined by simulating such hypothetical loading in ANSYS. Figure 8 illustrates the results of the disk displacement simulations. These results are approximated by:

$$u_F^{\circ}(t_e) = 0.38557 + 8.55627 \times 10^{-5} \cdot t_e + 1.83458 \times 10^{-8} \cdot t_e^2 \quad (19)$$

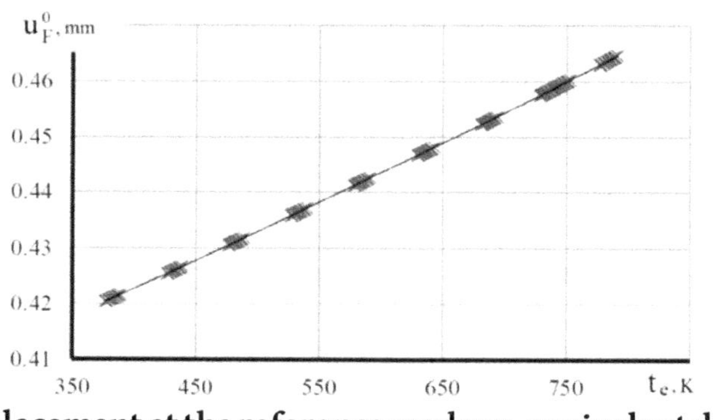

Figure 8. Disk displacement at the reference mode vs. equivalent disk temperature.

6.2. Characteristic Temperature

As follows from Eqs. (10) and (12), thermal loading on each heated part (disk, blade, and casing) depends on the temperature THPC (temperature of HPC air) and the similarity coefficients kT and k . As described in Section 4, the radial displacement uF caused by the force depends on the temperature state of HP and, therefore, is related to the thermal loading. Thus, this relation can be written by a function uF=f(THPC,kT,k) . The three interrelated arguments make this function complex for realization. Paper [22] proposes the concept of a characteristic temperature to be used as the unique function argument. The characteristic temperature ˜T is defined as a weighted mean of a boundary temperature T.

$$\tilde{T} = \frac{\int_A T(A)\alpha(A)dA}{\int_A \alpha(A)dA} \qquad (20)$$

where A is the surface of a heated part.

The characteristic temperature $\tilde{T}$ has an important property that temperatures t of a heated part tend to a value $\tilde{T}$ when heat transfer approaches zero, i.e.:

$$\lim_{k_\alpha \to 0} t = \tilde{T} \qquad (21)$$

This property allows us to determine the characteristic temperature through ANSYS simulation of the heated part with an extremely low similarity coefficient k. It is proven that such simulation yields low errors relatively a direct calculation of $\tilde{T}$ according to Eq. (20). For example, given k ≈10-3, the error was 0.01 K.

The characteristic temperature was firstly computed at the reference mode and, with the known value $\tilde{T}$0, a temperature coefficient:

$$\tilde{\Theta} = \frac{\tilde{T}^0 - T^0_{HPC}}{T^0_g - T^0_{HPC}} \qquad (22)$$

was formed, where Tg denotes a HPT input temperature. Then, it was found that this coefficient does not depend on an operating mode and can be used to determine the characteristic temperature at any mode by a simple relation:

$$\tilde{T} = \tilde{\Theta} \cdot (T_g - T_{HPC}) + T_{HPC} \qquad (23)$$

6.3. Static Force-Induced Displacement

To determine the relation between the temperatures $\tilde{T}$ and te, series of simulations in ANSYS have been conducted. For the disk under reference mechanical load, the thermal load parameters THPC, kt, and k are varied and the displacement u0F=f(THPC,kt,k) was determined for each combination of THPC, kt , and k . The equivalent temperature te corresponding to each displacement was found from Eq. (19). The characteristic temperature $\tilde{T}$ was calculated according to Eq. (23) using a known value THPC and a gas temperature Tg computed by NDM. By doing so, multiple pairs of te and $\tilde{T}$ values were found. With these data, the relation is between te and $\tilde{T}$ is described by:

$$t_e = -16.631 + 1.0518 \cdot \tilde{T} - 3.9362 \times 10^{-5} \cdot \tilde{T}^2 \qquad (24)$$

Thus, through a consecutive application of Eqs. (23), (24), (19), and (17), we can calculate a force-induced radial displacements of the disk as a function of the gas path variable THPC and Tg computed by NDM. The displacements of this enhanced algorithm as well as the original algorithm that consider constant disk elasticity were estimated by the comparison with the results of ANSYS-based simulations.

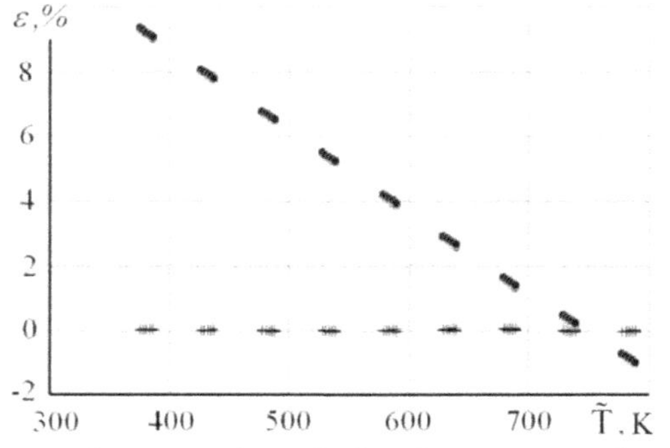

Figure 9. Errors of two force-induced displacement algorithms (scored line: algorithm that considers temperature-dependent elasticity; dashed line: algorithm that uses constant elasticity).

Figure 9 presents the errors of both algorithms for different characteristic temperatures and consequently for different engine operating points. We can see that the original algorithm has significant errors (up to 9%), whereas for the enhanced algorithm, the errors are negligible (within 0.1%). So, the accuracy of the displacement simulation was drastically enhanced despite the simplicity of the proposed algorithm.

6.4. Dynamic Force-Induced Displacement

The equivalent temperature te determined in Eq. (24) as a function of $\tilde{T}$ corresponds to a completely warmed-up heated part and its final static displacement. Let us call this temperature a static equivalent temperature tste($\tilde{T}$) . When the boundary conditions have changed, the force-induced displacements will vary dynamically and the temperature te will dynamically approach tste. As the relation between the displacement and the temperature te is practically linear (see Figure 8), their dynamic behavior will be similar. For this reason, the dynamics of te are described using the same displacement transient performances presented in Figures 5 and 6. For the disk, the algorithm to compute te is similar to that described in Section 5 for the thermal expansion displacements. The resulting equations to compute the equivalent temperature:

$$t_{ej}(\tau, k_\alpha) = \frac{T_j(k_\alpha)}{T_j(k_\alpha) + \Delta\tau} t_{ej}(\tau - \Delta\tau) + \frac{\Delta\tau}{T_j(k_\alpha) + \Delta\tau} t_e^{st}(k_\alpha), j = 1, 2 \quad (25)$$

and

$$t_e(\tau, k_\alpha) = \sum_{j=1}^{2} k_j(k_\alpha) t_{ej}(\tau, k_\alpha), j = 1, 2 \quad (26)$$

are also similar to displacement Eqs. (15) and (16) and the same parameters Tj and kj are employed. Using the dynamic value te(τ,k) from Eq. (26) as an argument, a dynamic displacement u∘F(te) is determined from Eq. (19) and a total force-induced disk displacement uF from Eq. (17). The blade and casing force-induced displacements are computed by similar algorithms. All these algorithms correspond to Blocks 1.4 and 2.4 of the engine ENDM presented in Figure 1.

7. VERIFICATION OF THE ENHANCED NONLINEAR DYNAMIC MODEL

7.1. Verification of the Simplified Dynamic Clearance Model

To verify the simplified dynamic clearance model (see Section 2), the following engine dynamics test case was prepared:

during the time interval τ=0...120s , the turbofan engine operates at idle conditions (k =0.2031, kt =0.4125, kn=0.5929) to warm up turbine parts;

during the time interval τ=120...500s , engine operates under the reference mode conditions (k = 1.0, kt = 1.0, kn = 1.0).

Total HPT disk displacements (mechanical and temperature-induced) were computed for this test case in ANSYS and by the proposed SDCM (see Section 2). As shown in Figure 10, the simulation curves practically coincide. The maximum difference observed at the mode change moment is about 0.05 mm and then it lessens. Thus, the simplified model can be considered accurate enough and can be used within ENDM of the turbofan engine under analysis.

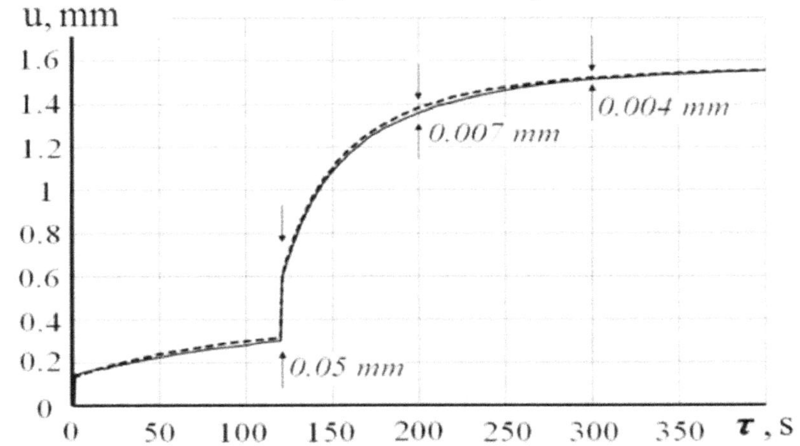

Figure 10. Total disk displacement simulation (dashed line—ANSYS; solid line—SDCM).

7.2. Accuracy of the Simulation of Engine Dynamic Performance

As mentioned in the beginning of Section 2, the enhanced nonlinear dynamic model (ENDM) has been developed for a turbofan engine of a maneuverable aircraft. The main objective was to help with the synthesis and adjustment of the algorithms of an engine automatic control system. The developed ENDM is based on the original nonlinear dynamic model (NDM) and the simplified dynamic clearance models (SDCMs) created for a high-pressure turbine (HPT) and a low-pressure turbine.

To verify the accuracy of the ENDM, it was compared with original NDM and with experimental data. A test-case transient was set by a low-pressure rotor speed nLP (control variable) profile and constant ambient conditions.

The profile presents a constant speed value 8100 rpm during the first 175 s, than a linear change to 12,400 rpm during 12 s, and the same constant value up to the transient end.

Figure 11 illustrates the dynamics of the HPT radial clearance simulated by ENDM in comparison with the steady-state clearance simulation (completely warmed-up turbine parts). One can state that ENDM correctly reflects the physics of real warming-up. From the beginning of the engine acceleration, the clearance descends in 15 s because the blade is rapidly warmed up. Next, the clearance grows due to the casing warming up. Finally, the clearance descends once more as the disk begins to warm up.

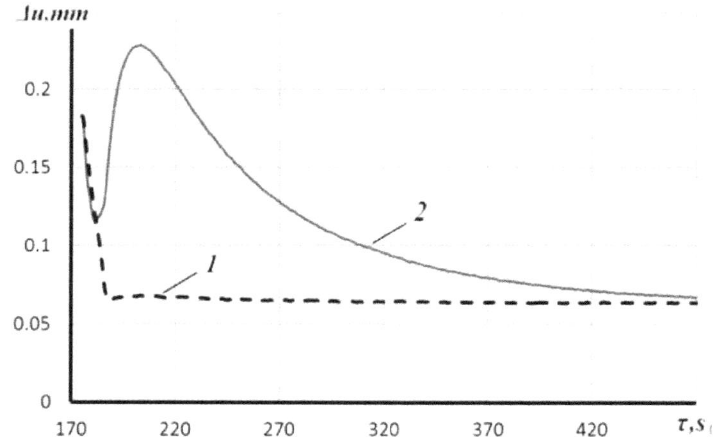

Figure 11. Dynamics of the HPT radial clearance (1—steady-state operating modes; 2—ENDM).

Figures 12 and 13 present the results of the comparison of the initial and enhanced dynamic models between each other and with experimental data for the same test-case transient. The plots of a fuel consumption variable in Figure 12 clearly show that the ENDM and experimental curves practically coincide. Both show the same fuel consumption overshoot after the control parameter change, and this overshoot gradually decreases during 150 s for both curves.

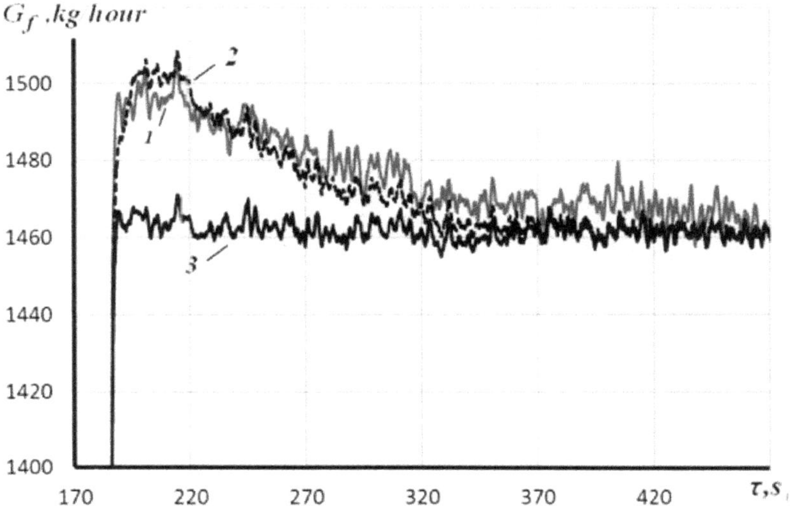

Figure 12. Fuel consumption dynamics (1—experimental data; 2—ENDM; 3—NDM).

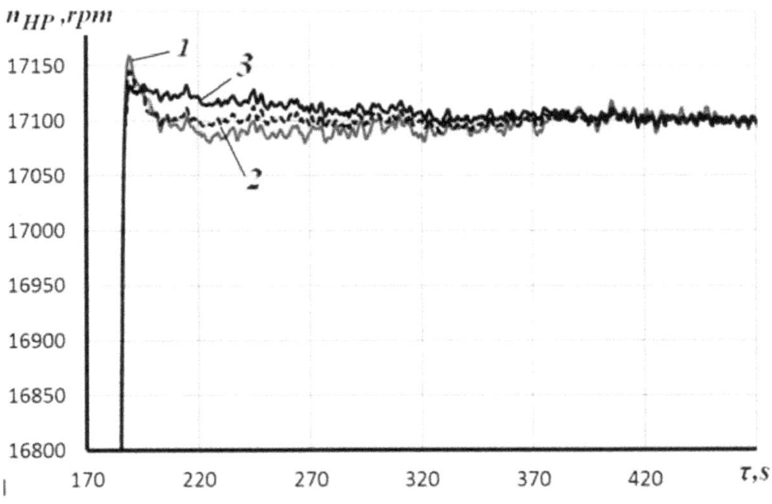

Figure 13. High-pressure rotor speed dynamics (1—experimental data; 2—ENDM; 3—NDM).

This elevated fuel consumption is explained by increased turbine clearances due to the delay in disk warming-up. In contrast, the NDM curve does not manifest a visible overshoot, and the transient process is by far shorter. One can make the same conclusion analyzing the plots of a high-pressure rotor speed in Figure 13:

the ENDM curve better fits experimental data, in particular, better reflects the effect of increased clearances.

The thrust is the principal parameter of a turbofan. However, under the control law nLP = const used in the experiments, it is constant as well, and the increased clearances are compensated by the additional fuel consumption observed in Figure 12.

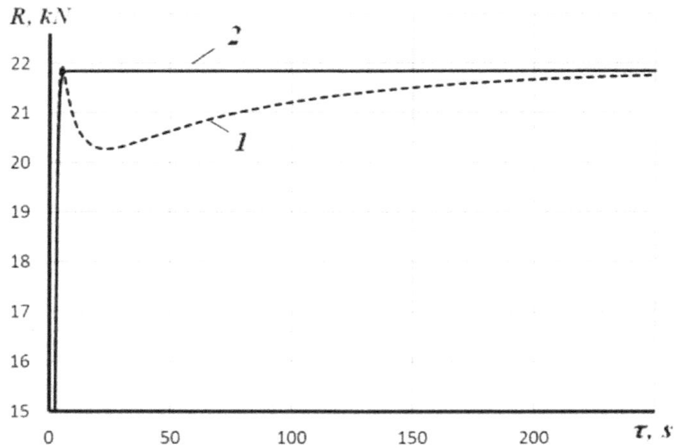

Figure 14. Thrust dynamics (1—ENDM; 2—NDM).

The thrust is the principal parameter of a turbofan. However, under the control law nLP = const used in the experiments, it is constant as well, and the increased clearances are compensated by the additional fuel consumption observed in Figure 12.

To show the impact of the clearances dynamically changed on the thrust, the simulation of the turbofan under the control law of a constant low pressure turbine temperature was performed. Figure 14 shows the thrust simulated by both models. It can be seen that, during the first 5 s of intensive engine dynamics, both models are equal.

Then, the NDM thrust remains constant, whereas the ENDM thrust begins to decrease with a maximal 7% thrust dip at the 12th second. Finally, the thrust gradually increases up to a steady-state value. Such behavior of the thrust simulated by ENDM completely corresponds to the known empirical information about the clearance influence.

To show the impact of the clearances dynamically changed on the thrust, the simulation of the turbofan under the control law of a constant low pressure turbine temperature was performed. Figure 14 shows the thrust simulated by both models. It can be seen that, during the first 5 s of intensive engine dynamics, both models are equal.

Then, the NDM thrust remains constant, whereas the ENDM thrust begins to decrease with a maximal 7% thrust dip at the 12th second. Finally, the thrust gradually increases up to a steady-state value. Such behavior of the thrust simulated by ENDM completely corresponds to the known empirical information about the clearance influence.

In this way, all the comparison results show that, first, the dynamic clearance influence is significant and cannot be neglected, and, second, the enhanced nonlinear dynamic model accurately simulates this effect and in general provides by far more realistic simulation than the initial dynamic model does.

8. CONCISE SUMMARY

This chapter outlines a new approach to improve a comprehensive physics-based nonlinear gas turbine model commonly utilized for aircraft engine control and diagnostics. This strategy enables us to address the influence of different turbine tip clearances on the dynamic engine performance. This issue is particularly crucial for the propulsion systems of agile aircraft. An improved nonlinear dynamic model of a turbofan engine for a maneuverable aircraft was produced by enhancing an initial nonlinear dynamic model and a simplified dynamic clearance model based on the finite element simulation results of turbine parts. The comparison between the original model and experimental data verified a significant enhancement in the accuracy of dynamic gas turbine simulation.

9. AT A GLANCE

Gas turbine development need several mathematical models. Physics-based nonlinear dynamic models are commonly utilized in creating control and diagnostic systems. This chapter initially examines existing studies on nonlinear dynamic engine modeling, focusing on model applications and advancements. Modeling the heating of engine components is seen as a significant advancement. This phenomenon involves a radial clearance alteration during transients that affects the engine's static and dynamic performances. The clearance change is typically calculated using a finite element method that is crucial for computing resources. The chapter introduces a faster simulation method to combine two dynamic processes: engine transients and clearance changes. This enables the development of a precise and efficient engine dynamic model that is user-friendly for designing control and diagnostic systems. The chapter discusses a methodology enhancement that involves considering how changes in metal temperature affect the stresses caused by mechanical loads. The methodology is validated by applying it to a specific turbofan engine and comparing the dynamic performances of the simulated and actual engine.

REFERENCES

1. Boyce MP. Gas Turbine Engineering Handbook. 4th ed. Oxford: Elsevier Inc.; 2012. 956 p. ISBN: 978-0-12-383842-1
2. Kulikov GG, Thompson HA. Dynamic Modelling of Gas Turbines: Identification, Simulation, Condition Monitoring, and Optimal Control. London: Springer; 2004. 309 p. ISBN: 1-85233-784-2
3. Asgari H, Chen XQ. Gas Turbines Modeling, Simulation and Control Using Artificial Neural Networks. New York: CRC Press, Tailor & Francis Group. 2016. 176 p. ISBN 13: 978-1-4987-2661-0
4. Saravanamuttoo HIH, Rogers GFC, Cohen H. Gas Turbine Theory. 5th ed. Edinburg: Person Education Limited; 2001. 491 p. ISBN 13: 978-0-13-015847-5
5. Saravanamuttoo HIH, Mac Isaac BD. Thermodynamic models for pipeline gas turbine diagnostics. ASME Journal of Engineering for Power. 1983;105:875-884
6. Stamatis A, Mathioudakis K, Papailiou KD. Adaptive simulation of gas turbine performance. Journal of Engineering for Gas Turbines and Power. 1990;112:168-175
7. Kamboukos P, Mathioudakis K. Multipoint non-linear method for enhanced component and sensor malfunction diagnosis. In: Proceedings of IGTI/ASME Turbo Expo; 8-11 May 2006. Barcelona, Spain; 2006. 9 p. ASME Paper GT2006-90451
8. Tsoutsanis E, Meskin N, Benammar M, Khorasani K. An efficient component map generation method for prediction of gas turbine performance. In: Proceedings of IGTI/ASME Turbo Expo; 16-20 June 2014. Dusseldorf, Germany; 2014. 12 p. ASME Paper GT2014-25753
9. Aretakis N, Roumeliotis I, Mathioudakis K. Performance model “zooming” for in-depth component fault diagnosis. Journal of Engineering for Gas Turbines and Power. 2011;133(3):031602
10. Kurzke J. Advanced user-friendly gas turbine performance calculations on a personal computer. In: Proceedings of ASME Gas Turbine Conference. 1995. ASME paper 95-GT-147
11. GasTurb 12. Manual; 2015. http://www.gasturb.de/Gtb12Manual/GasTurb12.pdf, last visited 16.12.2018
12. Volponi A, Brotherton T, Luppold R. Empirical tuning of on-board gas turbine engine model for real-time module performance estimation. In: Proceedings of IGTI/ASME Turbo Expo; 14-17 May 2007. Montreal, Canada; 2007. 10 p. ASME Paper GT2007-27535

13. Jaw LC, Mattingly JD. Aircraft Engines Controls: Design, System Analysis, and Health Monitoring. Reston, Virginia: American Institute of Aeronautics and Astronautics, Inc.; 2009

14. Litt JS et al. A Survey of Intelligent Control and Health Management Technologies for Aircraft Propulsion Systems. NASA Report TM-2005-213622; 2005. 21 p

15. Sobey AJ, Suggs AM. Control of Aircraft and Missile Power Plants. New York: Wiley; 1963

16. Gritsenko EA, Danilchenko VP, Lukachev SV, et al. Some Issues of the Design of Aircraft Gas Turbine Engines. Samara: Russian Federation: "СНЦ РАН"; 2002. 527 p. ISBN: 5-93424-057-9. (In Russian)

17. Kypuros JA, Melcher KJ. A Reduced Model for Prediction of Thermal and Rotational Effects on Turbine Tip Clearance. Tech. Rep. NASA. TM-2003-212226; 2003

18. Archipov AN, Karaban VV, Putchkov IV, et al. The whole-engine model for clearance evaluation. In: Proceedings of ASME Turbo Expo; 8-12 June 2009. Orlando, Florida, USA; 2009. ASME Paper GT2009-59259

19. Kurzke J. Transient simulations during preliminary conceptual engine design. In: Proceedings of XX International Symposium on Air Breathing Engines (ISABE 2011); 12-16 September 2011. Goethenburg, Sweden; 2011. ISABE-2011-1321

20. Merkler RS, Staudacher S. Modeling of heat transfer and clearance changes in transient performance calculations—A comparison. In: Proceedings of ASME Turbo Expo; 8-11 May 2006. Barcelona, Spain; 2006. ASME Paper GT2006-90041

21. Yepifanov S, Zelenskyi R, Loboda I. Modeling the gas turbine engine under its dynamic heating conditions. Journal of Engineering for Gas Turbines and Power—Transactions of ASME. 2015;137(3):1-10. Paper 031506. ISSN 0742-4795

22. Zelenskyi R, Yepifanov S, Martseniuk Y, et al. Dynamic turbine clearance simulation considering the influence of temperature on mechanical load-induced displacements. Journal of Aerospace Engineering. 2017;30(5):1-11. DOI: 10.1061/(ASCE)AS.1943-5525.0000751

23. Bouillet P. L'evolution de la technology des turboreactours de forte puissance. Aeronautique. 1984;107:4-29

24. Huebner KH, Dewhirst DL, Smith DE, et al. The Finite Element Method for Engineers. 4th ed. USA: John Wiley & sons, Inc; 2001

25. Oleynik OV. The concept and methods of lifetime depletion monitoring of gas turbine air-engine based on a dynamic identification of thermal and stress condition of main details [Ph.D. thesis]. Kharkov, Ukraine: National Aerospace University; 2006

8 Chapter

IMPACT OF MICROSTRUCTURE ON MICROHARDNESS AND ELECTROCHEMICAL PERFORMANCE IN HYPEREUTECTIC AL-FE ALLOY TREATED WITH LASER SURFACE REMELTING

1. INTRODUCTION: AN OVERVIEW

Laser surface remelting (LSR) has gained attention recently due of its unique characteristics. The high energy density of LSR allows for efficient energy utilization in remelting by altering the surface properties of a material while preserving its bulk properties. Laser Surface Remelting (LSR) causes the quick cooling of the molten material through conduction into the cold subsurface following rapid irradiation. Kalita [1] reported similar phenomena when utilizing laser surface melting (LSM) technology to investigate high strength aluminum alloys (HSAL). Pariona et al. [2, 3] employed the LSR approach in their investigation of a hypoeutectic Al-1.5 wt.% Fe alloy. The analysis of the cast region showed the development of a well-defined, compact, and very uniform microstructure, along with cracking, particularly with a significant increase in the production of protrusions on the weld joints compared to the untreated alloy. An overlapping line of sequential weld fillets was seen in the cast zone of this alloy, leading to a 61% increase in hardness relative to the base material. Pariona et al. [4] state that the Marangoni effect impacts the thermal gradient in the molten pool at high temperatures. This effect also influences the quality, features of microstructure, morphological characteristics, and the quality of the laser-treated workpiece track. However, these scientists found that at low laser beam velocities, the morphology is more pronounced and the quality of the track shows numerous faults compared to high laser beam velocities. Trdan and Grum [5] found that the laser shock peening (LSP) method can enhance corrosion resistance by raising the pitting potential and reducing the severity of pitting on the surface of the specimen. Hatamleh et al. [6] verified the increased corrosion resistance of laser-peened friction stir-welded 7075 aluminum joints when exposed to a 3.5% NaCl solution. Pariona and Micene [7] and Pariona et al. [2] examined the melted zone following LSR-treatment in an Al alloy. Their LAXRD study showed that the zone consisted mostly of metastable phases, with Al2O3 and AlN phases being the most prevalent. The authors highlighted how these phases contributed to the microstructural change, enhancing the high hardness and corrosion resistance of the workpiece treated with LSR in sulfuric acid. This work utilized liquid state route (LSR) treatment on hypereutectic Al-2.0 wt.% Fe alloy. The materials were analyzed using optical microscopy (OM), scanning electron microscopy (SEM), and Vickers microhardness test. Vickers hardness analysis was conducted on the cross-sectional area and surface of the treated sample. The electrochemical impedance spectroscopy (EIS) test was examined, and a numerical simulation was conducted. An analysis was conducted on the microstructure, microhardness, and electrochemical behavior of the laser-treated layer to establish a connection between these attributes and the procedure used.

2. MATERIALS AND METHODS

2.1. Material

Hypereutectic Al-2.0 wt.% Fe alloy under study was prepared with commercially pure raw materials. The material was cast in a resistance furnace (muffle) by pouring the liquid metal into a cylindrical ingot mold and cooling in ascending mode. Resulting ingot was sectioned into various samples, which were sand blasted individually to determine the chemical composition of alloy by energy-dispersive X-ray fluorescence spectrometry (Shimadzu EDX-7000), as indicated in Table 1.

Table 1. Chemical composition of materials used for manufacture of Al-2.0 wt.% Fe alloy.

Material	Impurity			
	Fe	Si	Cu	Ni
Al 99.76%	0.09%	0.06%	0.06%	0.03%
Fe 99.97%	—	0.01%	0.01%	0.01%

2.2. Laser Surface Treatment

In this research, Al-2.0 wt.% Fe alloy was subjected to laser surface remelting (LSR), without gas protection, with the purpose of generating metastable phases, using a 2 kW Yb-fiber laser (IPG YLR-2000S) in order to examine treated and untreated layers. LSR treatment was performed in a laboratory at Institute for Advanced Studies (IEAv) of *Aerospace Technical Center* (CTA–ITA) in São Jose dos Campos, SP, Brazil. A laser scanning speed of 40 mm s-1 was applied.

Average power of the laser beam was set at 600 W and the power density on the sample surface was estimated at 4.8 × 105 W cm−2. Laser-treated samples were covered with several weld fillets during the remelting process [8].

2.3. Equipment for Microstructural and Morphological Characterization

Various microstructural characterization techniques were employed to gain a better understanding of microstructural effects of Al-2.0 wt.% Fe alloy LSR-treated under study. These techniques applied were optical microscopy (OM), field-emission scanning electron microscopy (FESEM) coupled to energy dispersive spectroscopy (EDS) and Vickers microhardness testing, which are described in detail below.

LSR treated samples were analyzed by OM (Olympus BX51) couple to a Q-Color 3 digital camera to capture images. Prior to studying the LSR treated layer, the cross-sections were cut of the samples using a diamond blade and they were sanded and polished. Samples were chemically etched with hydrofluoric acid 0.5% (v/v) at intervals 30 to 45 seconds, after they were polished with metallographic polishing pads, using only water, to ensure LSR treatment would not be impaired. Laser-treated material and substrate were analyzed by FESEM (MIRA 3 LM) coupled to EDS to examine the microstructural changes caused by laser treatment.

2.4. Vickers Microhardness Testing

Vickers hardness (HV) tests were performed using a Leica VMHT MOT microhardness tester operating with a load of 0.1 kg at 15 seconds (HV 0.1 15 s). The tester was applied in the cross-sectional area of treated specimen, to different penetration depths until it reached the base material. Penetration depths of the tester from the surface in the treated material region were approximately 50, 100 and 200 μm, however, 300, 500 and 700 μm were in the base material region, as shown schematic in Figure 1. At each of these depths, 15 micro-indentations were made in lines parallel to surface. Average hardness and standard deviation at each of selected depths were calculated based on data obtained.

For preparation of HV tests, a cross-sectional sample was sanded with 600 and 1200 grit sandpaper and polished with colloidal silica to reduce its roughness, thereby preventing roughness that could interfering in results of HV measurements. Besides, microhardness was measured on the laser-treated sample surface, which was cleaned only with water to prevent that it could be modified. Furthermore, the material's hardness was tested on the weld fillets region and between them.

2.5. Electrochemical Impedance Spectroscopy (EIS) Test

The electrochemical impedance spectroscopy (EIS) test was performed in aerated solution of 0.1 M H2SO4 at a temperature of 25 ± 0.5°C, using Autolab PGSTAT 30 potentiostat system connected to a microcomputer.

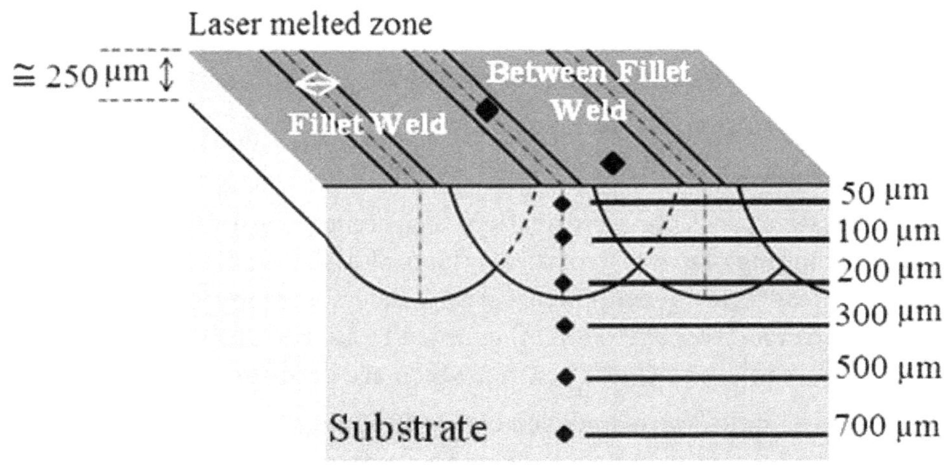

Figure 1. Schematic diagram of weld fillets on the sample surface and in the cross-sectional area showing the penetration depth of Vickers indenter in LSR-treated sample.

Working electrodes of surface-treated and untreated samples were prepared with epoxy resin to expose a top surface.

3. RESULTS AND DISCUSSION

3.1. Surface Characterization of Laser-Treated Samples

Figure 2 illustrates the morphology of hypereutectic Al-2.0 wt.% Fe alloy laser-treated analyzed by OM and FESEM, showing characteristics of the weld fillets formed during laser treatment. OM image in Figure 2(a) shows the surface morphology, while FESEM image in Figure 2(b) shows the morphology in more detail on the weld fillets region and between the weld fillets. As can be seen, on the weld fillet region contains a higher concentration of defects than between the weld fillets region. Zhang et al. [9] and Kalita et al. [1] reported a similar result. In Figure 2(b), the distance between the weld fillets is approximately 300 µm. Note the presence of several nanopores, which may be attributed to volatilization of inclusions or vaporization of the substrate itself, caused by hydrogen and moisture in the atmospheric air, which are absorbed in the laser-treated region, favoring the formation of pores. These results are consistent with reported of Yilbas et al. [10] and Pariona et al. [2]. The micrograph in Figure 2(c) shows on the weld fillets region under higher magnification, showing concentration of defects in more detail. Figure 2(d), also at increased magnification, shows between the weld fillets region, revealing a more uniform morphology with a columnar-like structure. Pariona et al. [2] also observed these structures in Al-1.5 wt.% Fe and Li et al. [11], these last authors stated that Al-Co-Ce alloys contain Al-rich eutectic regions whose structure and was similar to Al-2.0 wt.% Fe alloy. Peculiar characteristics of the microstructure shown in Figure 2(d), so it presented highly improved properties, such as: hardness, corrosion and wear resistance, which is resulted of precipitates dissolution and formation of metastable phases, to respect, several authors have reported similar results, among them, Damborenea [12], Pinto [13], Yue et al. [14], Majumdar et al. [15], Bertelli et al. [16], and Pariona et al. [2].

Pariona et al. [2] analyzed hypoeutectic Al-1.5 wt.% Fe alloy LSR-treated and observed presence of microcracks between the weld fillets. However, this phenomenon in this study was not observed in hypereutectic Al-2.0 wt.% Fe alloy LSR-treated, as can be seen in Figure 2(c) and (d).

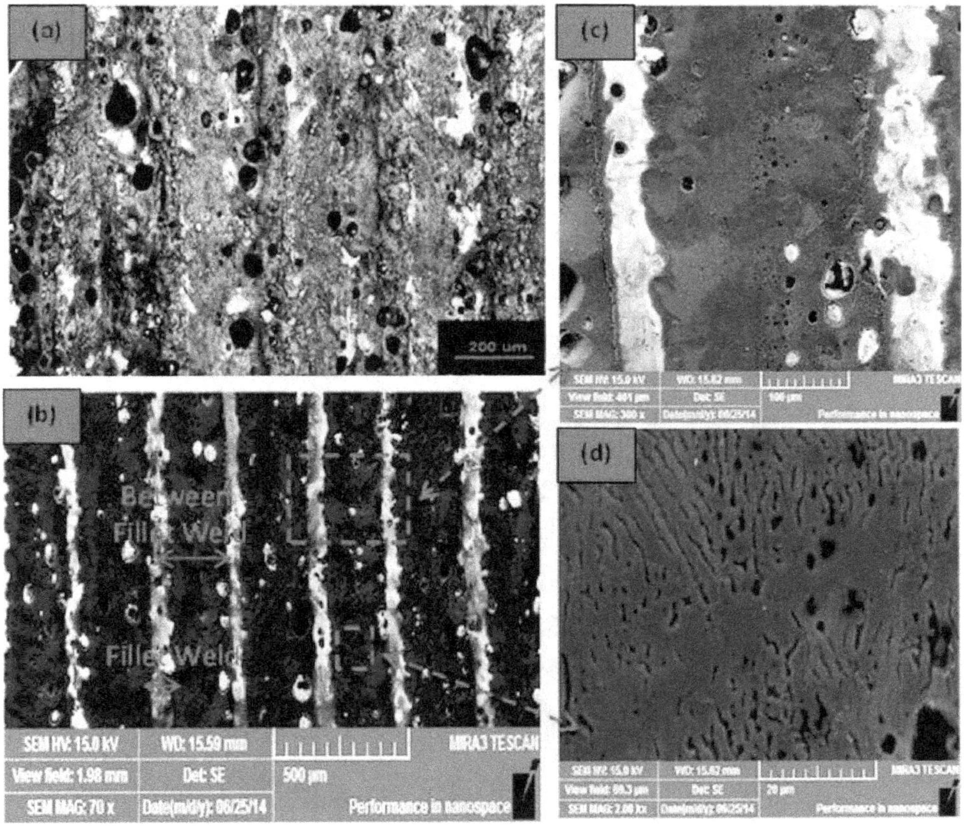

Figure 2. (a) OM, and (b) FESEM images of the morphology of hypereutectic Al-2.0 wt.% Fe alloy LSR-treated surface, showing regions on the weld fillet and between the weld fillets, (c) on the weld fillet region at increase magnification, and (d) between weld fillets region under higher magnification.

Absence of microcrack was expected, because, according to Mondolfo [17], formation Al-Fe alloys is impaired, when the material contains coarse Al3Fe particles or intermetallic phase, which tend to produce microcracks and reduce formability, whereas, this does not occur with presence of Al6Fe finely dispersed in Al-2.0 wt.% Fe alloy, however, the Al3Fe intermetallic phase does not appear in this alloy, as demonstrated by Pariona and Micene [7] by low-angle X-Ray diffraction analysis. Meanwhile, Gremaud et al. [18] reported, increasing the cooling rate of hypereutectic alloys containing up to 9 wt.% of Fe suppresses formation of stable Al3Fe phase, which is replaced by Al6Fe phase, which confirms our result.

3.2. Characterization in the Cross Section of Laser-Treated and Untreated Materials

Figure 3 shows the cross-sectional analysis by OM. In this region can be observed the penetration depth of the treated region was around 250 μm, and the distance between the weld fillets was approximately 300 μm (also was shown in the first micrograph, Figure 2). Note clearly visible difference of the treated region microstructure and of the substrate.

The laser melted surface micrograph is shown at Figure 3, as can be seen it is free of microcracks and the melted regions are free of precipitates too. Fine microstructure of the melt zone is attributed to high cooling rate. Microstructure obtained in this work is similar to other laser melted aluminum alloys reported in the literature, i.e., Watkins et al. [19] reported that the microstructure of laser melted AA 2014 consists of columnar grains growing epitaxially from the substrate.

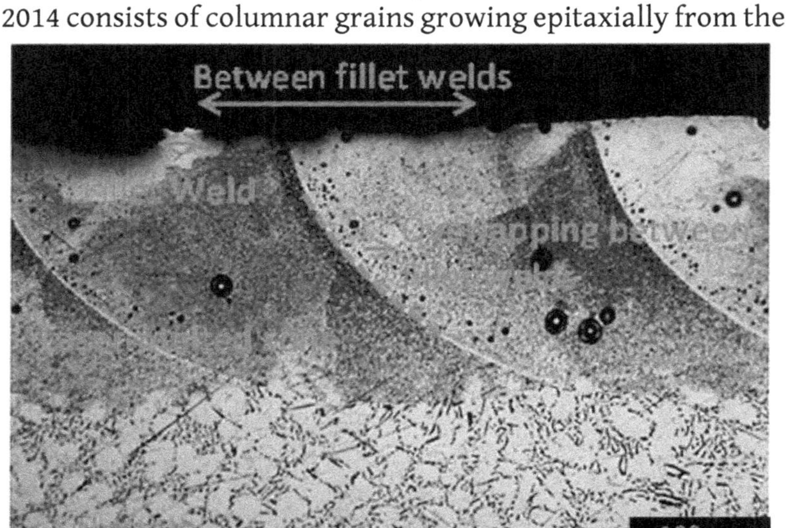

Figure 3. OM micrograph in the cross-sectional area of laser-treated material.

Although, maximum melt depth observed in this work was 250 μm (Figure 3); however the thickness of this zone depends of laser power and of the Marangoni effect, as was discussed by Pariona et al. [4, 8], these authors demonstrated when the laser beam velocity is low, therefore the molten zone depth is greater.

Figure 3 also shows zones where there is overlapping of consecutive weld fillets. This overlapping is more common in Al-2.0 wt.% Fe alloy than in Al-1.5 wt.% Fe alloy, reported by Pariona et al. [2, 3, 8]. Kalita et al. [1] also reported overlapping of consecutive weld fillets and Cordovilla et al. [20] pointed out as essential tool to understand way in which each track affects the microstructures produced by previous one.

Figure 4 depicts a cross-sectional LSR-treated sample and analyzed by SEM, showing some regions of substrate and the as-cast microstructure. In the cast area in Figure 4, note presence of protuberances, which correspond to on the weld fillet region (also shown in Figure 3). According to Pariona et al. [4], presence of protuberances is more noticeable in Al-1.5 wt.% Fe alloy than in Al-2.0 wt.% Fe alloy. Figure 4(a) also shows an overlapping line of consecutive weld fillets. Figure 4(b) and (e) show the substrate region and the laser-treated area under higher magnification, showing a visibly different microstructure, with a dendritic-like structure. This microstructural difference between untreated substrate and LSR-treated region is attributed to temperature applied on the material surface, which exceeded its melting point but was lower than boiling point, followed by rapid cooling in laser treatment process and this leads a high thermal gradient, and so in this way produces the laser melted zone.

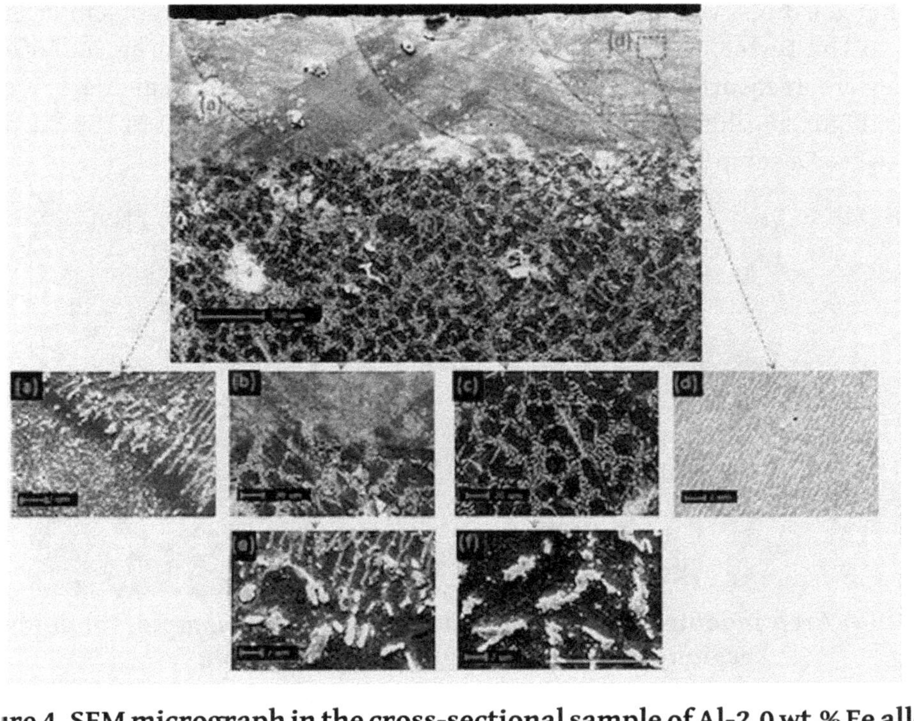

Figure 4. SEM micrograph in the cross-sectional sample of Al-2.0 wt.% Fe alloy LSR-treated: (a) overlapping line of consecutive weld fillets, (b) interface of treated surface and substrate, (c) substrate unaffected by laser treatment, (d) detail in the cast region, (e) interfacial region of the treated surface and substrate, and (f) detail of the substrate unaffected by laser treatment.

This treatment resulted in formation of a thin recast layer with a refined microstructure practically free of precipitates, inclusions and intermetallic phases [18], as can be clearly seen at the magnified image, Figure 4(d), with a columnar dendrite structure, Watkins et al. [19] and, Grum and Sturm [21] have also reported this characteristic in laser cast materials. Figure 4(c) shows the substrate region, which is also displayed under higher magnification in Figure 4(f), showing presence of intermetallic phase dispersed in the matrix. A comparison in more detail of Figure 4(d) and (f) reveals that the treated region morphology is more homogeneous, without presence of the intermetallic phase that extends throughout the recast area and showing evidence of transition from coarse-grained to fine-columnar-dendrite structure.

According to Pariona et al. [2], behavior of the laser treated region is homogeneous and similar to an amorphous phase; hence, it shows greater hardness, lower surface roughness, and higher corrosion resistance, reported by Pariona and Micene [7].

3.3. Vickers Microhardness Test

Vickers hardness test was accomplished in this work and by means of a microscope coupled to the tester, the "d1" and "d2" diagonals formed in area indented by pyramid were measured, and these parameters were used to calculate Vickers hardness. Figure 5 illustrates indented areas used for calculation of the hardness of Al-2.0 wt.% Fe samples.

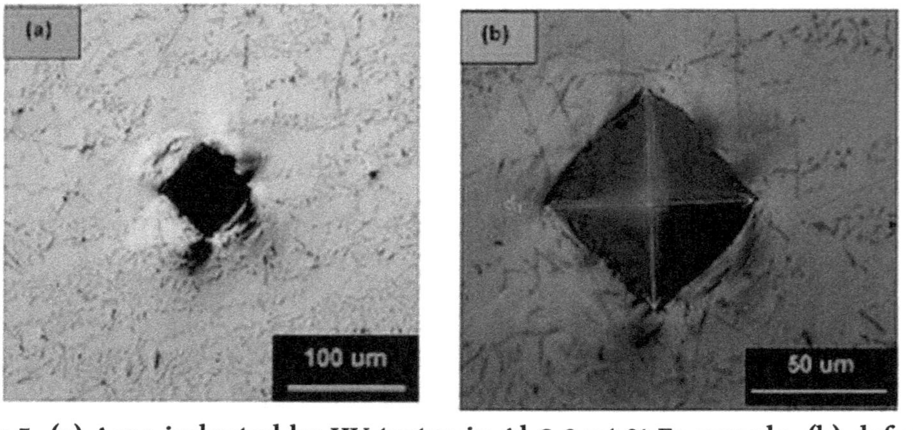

Figure 5. (a) Area indented by HV tester in Al-2.0 wt.% Fe sample, (b) deformed region shown under higher magnification.

Microhardness profiles were measured along in a cross-sectional sample, for laser-treated layer and untreated. These measurements were taken along lines parallel to surface at depths of 50, 100, 200, 300, 500 and 700 μm, applying a load of HV 100 gf for 15 s. Figure 6 illustrates the 15 micro-indentations made in the cross section at each of these depths to measure the hardness. Average hardness values and standard deviation (s.d.) at each depth were calculated based on these measurements, and are given in Table 2.

An analysis of the data in Table 2 indicates the HV is higher for the LSR treated region than the untreated region. The average hardness of the treated region is 58.8 HV, while that of the untreated region is 35.7 HV, which corresponds at 60.7% increase in hardness in the treated region compared to the untreated region.

The data in Table 2, also is shown in graphical form in Figure 7, it clearly show increase in hardness at treated region than untreated substrate. This difference is attributed to microstructural changes as resulting of LSR-treated.

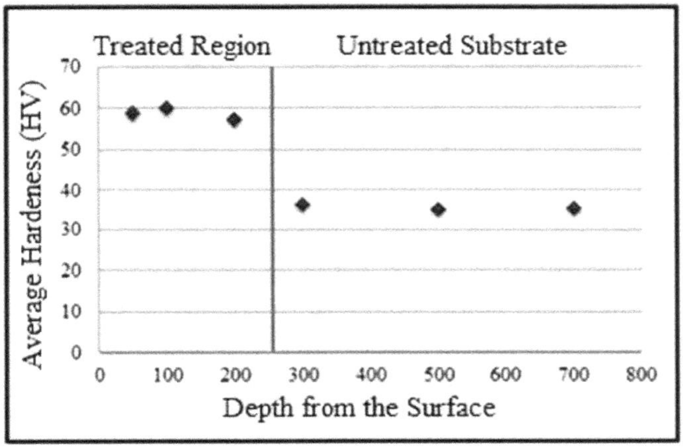

Figure 6. Vickers hardness analysis (HV 100 gf, 15 s) of LSR-treated layer and untreated substrate.

In other studies involving LSR treatment of materials, similar results have been obtained by Yao et al. [19] and others, who reported a significant increase in hardness in laser-treated region than untreated region.

Table 2.Vickers hardness analysis in a cross-sectional area, in sample treated and untreated (HV 100 gf, 15 s).

Region	Depth of the surface	Average of VH	Standard deviation of VH
Treated region	50 μmμm	59.0	3.15
	100 μmμm	60.0	3.8
	200 μmμm	57.4	3.0
Untreated region	300 μmμm	36.5	1.43
	500 μmμm	35.2	1.44
	700 μmμm	35.4	1.68

The material surface hardness was also analyzed by HV measurements on the weld fillets region and between them (see Figures 2–4), for the as-received laser-treated sample.

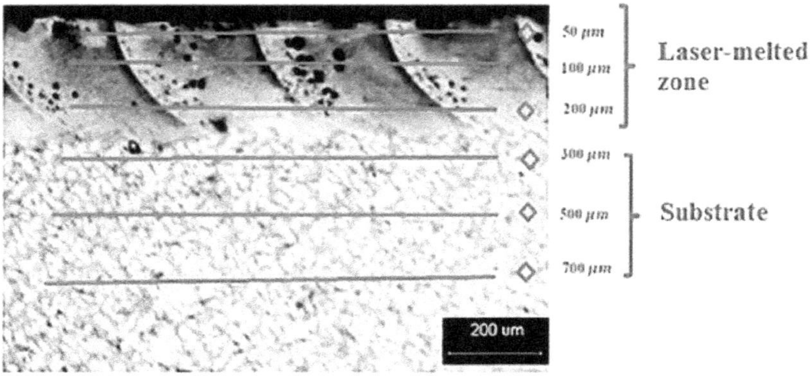

Figure 7. OM image in the cross section of Al-Fe sample laser-treated, indicating the depths selected for microhardness measurements.

As can be seen in Table 3, the HV values measured on the sample surface are consistent with those measured in the cross-section too, so showing a higher average hardness at the region between the weld fillets than on the weld fillet. Pariona et al. [3], who made a comparative analysis of the HV of Al-1.5 wt.% Fe alloy measured on the weld fillets and between the weld fillets, also reported that the hardness between the weld fillets was higher than on the weld fillets, therefore, the surface hardness in the laser-treated region in relation to the untreated region is high, due to the treated region morphology is more homogeneous, without presence of intermetallic phase (Al3Fe) and with the presence of Al6Fe phase finely dispersed in the matrix that extends throughout the recast area, as can be checked in Figures 2–4.

Table 3. Analysis of Vickers hardness on the treated sample surface, indicating the hardness at the regions on the weld fillets and between the weld fillets (VH 0.1 15 s).

Region	Average of HV	Standard deviation of VH
On the weld fillets	52.68	6.18
Between the weld fillets	59.14	5.53

The average Vickers hardness was calculated for 15 micro-indentations made on the weld fillets and between the weld fillets, as indicated in Table 3.

Present study focused on the microstructural characterization of hypereutectic Al-2.0 wt.% Fe alloy, while previous studies by Pariona et al. [2, 3, 4] involved hypoeutectic Al-1.5 wt.% Fe alloy. Although both alloys were castings and solidified by laser-treated process in the same conditions, however, microstructural analysis of the two alloys revealed characteristics different. The overlapping line of consecutive weld fillets at the cast zone of Al-1.5 wt.% Fe alloy was barely perceptible than Al-2.0 wt.% Fe alloy. In addition, also in the cast zone, presence of protuberances on the weld fillets was much more noticeable at Al-1.5 wt.% Fe alloy than at Al-2.0 wt.% Fe alloy. However, Al-1.5 wt.% Fe alloy showed a behavior lamellar at the cast zone and meanwhile Al-2.0 wt.% Fe alloy showed a behavior fine-columnar-like structure. Both alloys showed nanopores, which were concentrated mostly on the weld fillets. The microhardness of Al-2.0 wt.% Fe alloy LSR-treated surface was slightly more higher than Al-1.5 wt.% Fe alloy.

3.4. Influence of Microstructure on Electrochemical Behavior

Electrochemical impedance spectroscopy (EIS) after a longer immersion time of 3300 s (Ecorr) was also carried out. EIS experiments were performed at open circuit potential over a frequency range of 0.1–100 kHz. The sinusoidal potential perturbation was 5 mV in amplitude. Figure 8 shows Nyquist plots for untreated and LSR-treated alloys in aerated 0.1 M H2SO4 at a temperature of 25°C ± 0.5°C, after 3300 s immersion.

Trdan and Grum [5] pointed out, which EIS technique is probably one of the most powerful nondestructive steady-state methods in electro-chemistry. EIS enables us to determine different parameters of equivalent electrochemical systems (capacitance, resistance, electrolyte interface, etc.). Moreover, Kendig et al. [22] suggested that EIS spectra obtained over a wide range of frequencies indicate that the technique is right choice, since it is applicable for evaluating complicated corrosion processes.

By analyzing the diagram of Figure 8, presence of capacitive loops is observed at high frequencies and inductive loops at low frequencies. However, it is seen that LSR-treated workpiece exhibit higher resistance compared to untreated (as received) workpiece, at all immersion times, this result is in agreement with result studied by Trdan et al. [5, 23]. Capacitive behavior of material is related to aluminum oxide layer properties, studied by Pariona et al. [7], while inductive behavior can be attributed with active state of aluminum surface present in studied electrolyte. Passive regions refer to the oxide layer on aluminum, in accordance with Zhang et al. [24] and Pariona and Micene [7] argued that LSR-treated sample resulted in reduction of current density, and this fact indicates a lower corrosion, therefore, LSR-treated workpiece showed clearly a wide passive zone.

Electrochemical parameters obtained for Al-2.0 wt.% Fe alloy LSR-treated and untreated are shown in Table 4. Where, Rs is the polarization resistance of M H2SO4 solution. Cdl and RP are attributed to the electric double layer capacitor and the resistance of passive layer, respectively. It can be seen that results obtained for Rp are very close to those obtained through of potentiodynamic polarization curves, investigated for same alloy by Pariona and Micene [7], thus indicating reliability of presented results.

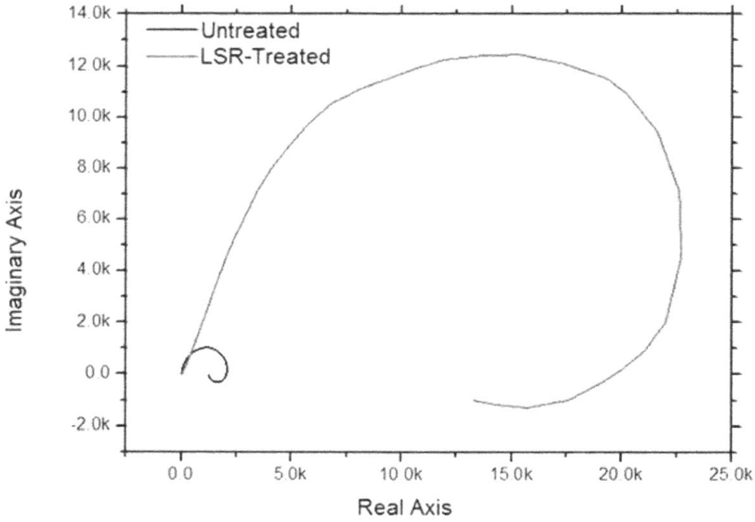

Figure 8. Nyquist plots for untreated and LSR-treated alloys in aerated 0.1 M H2SO4 at a temperature of 25°C.

It was also noted that LSR treatment provided the formation of thicker aluminum oxide when compared to untreated workpiece, argued by Pariona et al. [8]. According to result of Table 4, based on the values of Rp, LSR-treated layer presents greater resistance to charge transfer at electrode/solution interface in relation to untreated workpiece. Then, it can be emphasized that treated sample is more resistive than untreated workpiece, at 11 times higher. However, LSR-treated workpiece showed lower electric double layer capacitor values than untreated material. These results once again report that RSL treatment is an efficient technique to improve corrosion behavior of Al-2.0 wt.% Fe alloy in sulfuric acid medium, thus, Trdan and Grum [5] demonstrated improvement of corrosion resistance by means of increased pitting potential with lower intensity of pitting attack on the specimen's surface due to laser shock peening (LSP).

Table 4. Electrochemical parameters for Al-2.0 wt.% Fe alloy LSR-treated and untreated.

Material	R_s Polarization resistance	R_p Resistance of passive layer	C_{dl} Electric double layer capacitor
Treated	22 Ω	22.6 KΩ	1.94 μF
Untreated	22 Ω	2.07 KΩ	3.4 μF

There are several opinions and controversies of authors around pseudo-inductive that it presents themselves in EIS technique. According to Zhang et al. [24], the pseudo-inductive behavior was observed on microcapillaries. Silver et al. [25] argued, which in many cases, loops emerging in the low-frequency range are wrongly called inductive. In opinion of these last authors, the pseudo-inductive behaviors are caused by drift and corrosion and can be explained by so-called negative capacitance effect.

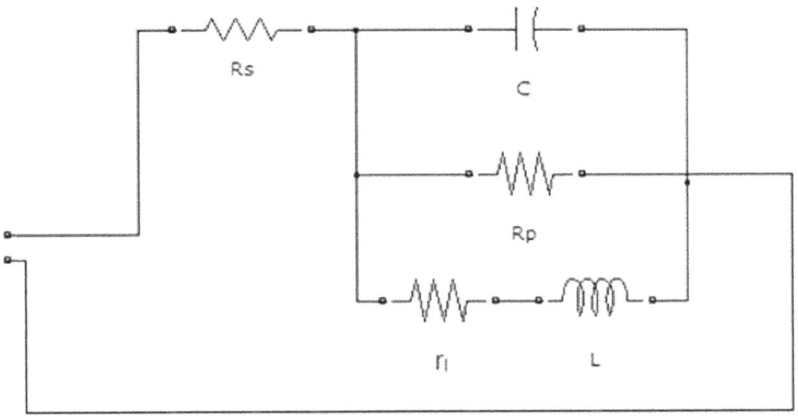

Figure 9. Electric behavior of treated layer. A proposed of the equivalent circuit.

According to impedance spectroscopy technique, it was possible to characterize electric behavior of treated layer and to design the values of equivalent circuit (EC) formed by resistors, capacitors and inductors. In Figure 9 were fitted a proposed of equivalent circuit, designed from Matlab software, the following the model suggested by Macdonald [26].

EC proposed (Figure 9) describe this system, where is given by Rs, which represents the solution resistance, Cdl is the electric double layer capacitor, Rp is the polarization resistance, R1 is the inductive resistance and L is the inductive element. The Rs, Rp and Cdl data were obtained experimentally by potentiostat software. However, L has no way to measure, for this a computational adjustment was made by Matlab software, using the circuit of Figure 9, then, for untreated sample, the corresponds values were found, for L = 1000 H and for treated sample was L = 5000 H.

Figures 10 and 11 show experimental results that were overlaid with simulated data for untreated and LSR-treated workpieces, respectively, thus, curves presented a good fit for the untreated case (Figure 10), where experimental and simulated values are very close.

Moreover, for LSR-treated case, curves did not present a good fit (Figure 11), experimental and simulated values are quite different. This is because, after LSR-treated, treated layer is composed of metastable phases, mainly consisting of alumina and aluminum nitride, besides, microstructure showing multiple laser tracks characteristics formed during laser treatment, however, the molten pool zone showed a fine microstructure due to high quenching rates applied, meanwhile,

Guan et al. [28] argued, which laser beam tracks has significant influence on surface quality of laser-treated materials. Consequently, LSR-treated layer has a very complex feature; with certainly, EC proposed should be more complex for LSR-treated.

Different authors reported several investigations about this study, which microstructure characteristic caused by overlapping ratios and multi-track, influence on electrochemical behavior or laser multi-track overlapping and consequently in effect of corrosion process. According to previous works, He et al. [27] point out that overlapping tracks affect heat transfer and liquid flow, microstructure evolution, further, Cordovilla et al. [20] pointed out as essential tool to understand way in which each track affects the microstructures produced by previous one. On the other hand, Guan et al. [28] argued that overlapping is important in determining corrosion resistance due to microstructure in-homogeneities in the molten pool.

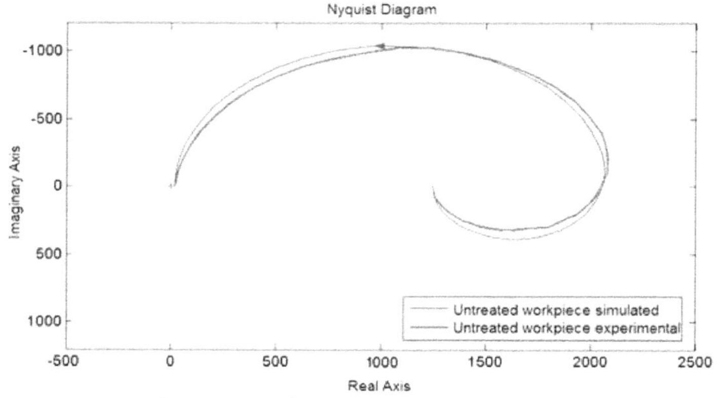

Figure 10. Nyquist diagram of experimental and simulated result of a untreated sample.

However, Kalita [1] noted existence of increase in corrosion resistance comes as a result of homogenization and microstructure refinement, which is due to the surface layer melting, as well as through decreases electrical conductivity of resultant passive layer, still these same authors argue, laser surface melting is a useful method for corrosion protection of friction stir weld surfaces as a result of improved microstructure and phase distribution. Nevertheless, Watkins et al. [19] reported, laser surface treatments offer significant potential for improvement of materials properties such as corrosion performance and wear resistance.

The authors Yue et al. [14] who reported, potentiodynamic polarization tests showed that as a result of laser treatment, the corrosion current can be reduced by as much as six times, and a passive region was obtained. Besides, analysis of electrochemical impedance measurements showed that at an open-circuit potential (OCP), the polarization resistance and double-layer capacitance of the film electrolyte interface of laser-treated specimen were one order of magnitude higher and six times lower than that untreated specimen, respectively.

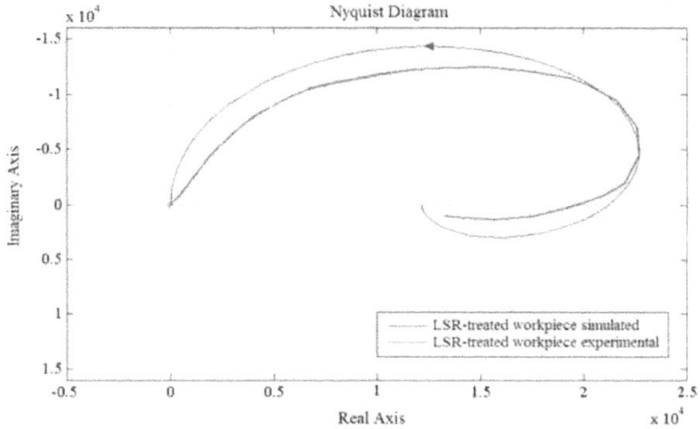

Figure 11. Nyquist diagram of experimental and simulated result of a LSR-treated workpiece.

Since then, hypereutectic Al-2.0 wt.% Fe alloy laser-treated is very peculiar and that it has very special characteristics. Therefore, in this study the influence of microstructural characteristic on microhardness and electrochemical behavior was demonstrated clearly, thus it has an innovative character and can be applied in aerospace, aeronautical and automobile industries. Guan et al. [28] argued, which overlapping adjacent traces as a result of multiple passes using scanning laser beam is usually adequate for production of area coverage. It has long been realized that laser beam overlapping may play a significant role in influencing final surface properties of laser-treated materials.

4. CONCISE SUMMARY

This research involved a study of hypoeutectic Al-2.0 wt.% Fe alloy subjected to a laser surface remelting (LSR) treatment. The main results are the following:

1. The cast region displays a refined, compact, and uniform microstructure without microcracks and with the creation of a minor protuberance.
2. The fine microstructure of the melt zone is a result of the high cooling rate from the LSR treatment.
3. The cast area of the Al-2.0 wt.% Fe alloy exhibited a distinct overlapping line of successive weld fillets.
4. The hardness of the cast region of Al-2.0 wt.% Fe alloy was approximately 61% greater than the untreated material.
5. Electrochemical impedance spectroscopy characteristics for Al-2.0 wt.% Fe alloy, both LSR-treated and untreated, exhibited capacitive loops at high frequencies and inductive loops at low frequencies.
6. Workpieces treated with LSR show an 11-fold increase in polarization resistance compared to untreated workpieces. The capacitive behavior of the material is influenced by the properties of the aluminum oxide layer.

7. The microstructure characteristics resulting from overlapping ratios and multi-track processes significantly impact the electrochemical process.
8. The impedance modeling of the proposed equivalent circuit for the Al-2.0 wt.% Fe alloy showed a close agreement between experimental and simulation results, with curves fitting well for the untreated scenario.

Conversely, in the case of LSR treatment, they were significantly distinct. This alloy has the potential to be used in the automotive, aerospace, and electronics industries because of its high hardness. The laser-treated alloy has a fine-columnar-like structure, unlike the untreated material.

5. AT A GLANCE

The microstructure of a hypereutectic Al-2.0 wt.% Fe alloy was analyzed using laser surface remelting (LSR), microhardness test, and electrochemical behavior test, followed by numerical simulation. The microstructure was examined using optical microscopy, field-emission scanning electron microscopy, and Vickers microhardness tests. The study found that the LSR treatment led to quick heating and subsequent rapid cooling, resulting in the creation of a thin recast layer with a more refined microstructure. This process also caused the dissolution of precipitates and inclusions, as well as the formation of metastable phases. However, an overlapping line was noticed on consecutive weld fillets. An investigation of Vickers hardness was conducted on the cross-sectional area and surface of the treated sample, resulting in higher microhardness in the treated region compared to the untreated substrate. The EIS test showed that workpieces treated with laser surface remelting have a polarization resistance 11 times higher than untreated workpieces. The capacitive behavior of the material is influenced by the properties of the aluminum oxide layer. Additionally, the microstructure characteristics resulting from the overlapping ratio and multi-track process significantly impact the electrochemical process.

REFERENCES

1. Kalita SJ. Microstructure and corrosion properties of diode laser melted friction stir weld of aluminum alloy 2024 T351. Applied Surface Science. 2011;257(9):3985-3997
2. Pariona MM, Teleginski V, dos Santos K, Machado S, Zara AJ, Zurba NK, et al. Yb-fiber laser beam effects on the surface modification of Al-Fe aerospace alloy obtaining fillet weld structures, low fine porosity and corrosion resistance. Surface and Coatings Technology. 2012;206:2293-2301
3. Pariona MM, Teleginski V, dos Santos K, dos Santos ELR, de Lima AAOC, Riva R. AFM study of the effects of laser surface remelting on the morphology of Al Fe aerospace alloys. Materials Characterization. 2012;74:64-76

4. Pariona MM, Taques AF, Woiciechowski LA. The Marangoni effect on microstructure properties and morphology of laser-treated Al-Fe alloy with single track by FEM: Varying the laser beam velocity. International Journal of Heat and Mass Transfer. 2018;119:10-19

5. Trdan U, Grum J. Evaluation of corrosion resistance of AA6082-T651 aluminium alloy after laser shock peening by means of cyclic polarisation and ElS methods. Corrosion Science. 2012;59:324-333

6. Hatamleh O, Singh PM, Garmestani H. Corrosion susceptibility of peened friction stir welded 7075 aluminum alloy joints. Corrosion Science. 2009;51:135-143

7. Pariona MM, Micene KT. The alumina film nanomorphology formed to improve the corrosion resistance of Al-2.0 wt.% Fe alloy as result of the laser surface melting technique applied. Advances in Chemical Engineering and Science (ACES). 2017;7:10-22

8. Pariona MM, Teleginski V, dos Santos K, de Lima AAOC, Zara AJ, Micene TM, et al. Influence of laser surface treated on the characterization and corrosion behavior of Al-Fe aerospace alloys. Applied Surface Science. 2013;276(2013):76-85

9. Zhang X, She J, Li S, Duan S, Zhou Y, Yu X, et al. Simulation on deforming progress and stress evolution during laser shock forming with finite element method. Journal of Materials Processing Technology. 2015;220:27-35

10. Yilbas BS, Al-Aqeeli N. Analytical investigation into laser pulse heating and thermal stresses. Optics & Laser Technology. 2009;41:132-139

11. Li R, Ferreira MGS, Almeida A, Vilar R, Watkins KG, McMahon MA, et al. Localized corrosion of laser surface melted 2024-T351 aluminum alloy. Surface and Coatings Technology. 1996;81:290-296

12. Damborenea J. Surface modification of metals by high power lasers. Surface and Coatings Technology. 1998;100–101:377-382

13. Pinto MAQ, Cheung N, Ierardi MCF, Garcia A. Microstructural and hardness investigation of an aluminum-cooper alloy processed by laser surface melting. Materials Characterization. 2003;50:249-253

14. Yue TM, Yan LJ, Chan CP, Dong CF, Man HC, Pang GKH. Excimer laser surface treatment of aluminum alloy AA7075 to improve corrosion resistance. Surface and Coatings Technology. 2004;179:158-164

15. Majumdar JD, Pinkerton A, Liu Z, Manna I, Li L. Microstructure characterization and process optimization of laser assisted rapid fabrication of 316L stainless steel. Applied Surface Science. 2005;247:320-327

16. Bertelli F, Meza ES, Goulart PR, Cheung N, Riva R, Garcia A. Laser remelting of Al-1.5 wt.% Fe alloy surfaces: Numerical and experimental analyses. Optics and Lasers in Engineering. 2011;49:490-497

17. Mondolfo LF. Aluminum Alloys: Structure and Properties. 9th ed. London: Butterworths; 1976

18. Gremaud M, Carrard M, Kurz W. The microstructure of rapidly solidified Al-Fe alloys subjected to laser surface treatment. Acta Metallurgia et Materialia. 1990;38:2587-2599

19. Watkins KG, Liu Z, McMahon M, Vilar R, Ferreira MGS. Influence of the overlapped area on the corrosion behaviour of laser treated aluminium alloys. Materials Science and Engineering A. 1998;252:292-300

20. Cordovilla F, García-Beltrán A, Sancho P, Domínguez J, Ruiz-de-Lara L, Ocaña JL. Numerical/experimental analysis of the laser surface hardening with overlapped tracks to design the configuration of the process for Cr-Mo steels. Materials and Design. 2016;102:225-237

21. Grum J, Sturm R. A new experimental technique for measuring strain and residual stresses during a laser remelting process. Journal of Materials Processing Technology. 2004;147:351-358

22. Kendig MW, Allen AT, Jeanjaquet SL, Mansfeld F. In: Baboian R, editor. Electrochemical Techniques. Houston, Texas, USA: NACE; 1986. pp. 67-71

23. Trdan U, Skarba M, Grum J. SEM/EDS characterization of laser shock peening effect on localized corrosion of Al alloy in a near natural chloride environment. Corrosion Science. 2014;82:328-338

24. Zhang DQ, Li J, Joo HG, Lee KY. Corrosion properties of Nd:YAG laser—GMA hybrid welded AA6061 Al alloy and its microstructure. Corrosion Science. 2009;51:1399-1404

25. Silver BR, Holub K, Marecek V. Low frequency pseudo-inductive phenomenon at novel glass microcapillaries exhibiting non-ohmic behavior. Electrochimica Acta. 2013;110:801-808

26. JR MACDONALD. Impedance Spectroscopy. 2nd ed. New York: Wiley Interscience; 2005

27. He X, Elmer JW, DebRoy T. Heat transfer and fluid flow in laser microwelding. Journal of Applied Physics. 2005;97(8):084909-084909-9

28. Guan YC, Zhou W, Li ZL, Zheng HY. Influence of overlapping tracks on microstructure evolution and corrosion behavior in laser-melt magnesium alloy. Materials and Design. 2013;52:452-458

9

Chapter

OPTIMAL CONTROL OF FUZZY SYSTEMS WITH APPLICATION TO RIGID BODY ATTITUDE CONTROL

1. INTRODUCTION: AN OVERVIEW

Lotfi Aliasker Zadeh invented the fuzzy set and fuzzy logic, which have been widely implemented in several disciplines on Earth and proven to be beneficial in various businesses [1,2,3]).

Fuzzy logic is a multi-valued logic system that permits the definition of intermediate values between traditional binary assessments such as yes or no, true or untrue, and big or little. Fuzzy concepts are characterized by their ability to address complex system problems through fuzzy and linguistic modeling. By incorporating fuzzy logic into system design, it is feasible to create knowledge-oriented intelligent systems. This is a crucial attribute of fuzzy logic.

Most physical systems are nearly nonlinear dynamic systems. Traditional control design methods utilize several approximation techniques such linear, piecewise linear, and lookup table approximations to address nonlinearities. The linear approximation method transforms a nonlinear dynamic system into a linearized design-model by focusing on a specific equilibrium point. The controller is created to meet a certain control purpose for the linearized design-model. The linear approximation method is straightforward but can constrain control performance. The controller created using the linear approximation method is only reliable when the states of a nonlinear dynamic system operate near the equilibrium point being considered. This is a fundamental constraint of control design through linear approximation. The piecewise linear approximation method necessitates the creation of multiple linear controllers. Therefore, it outperforms the linear approximation method, although being laborious to execute. The lookup table approximation method can enhance control performance, although it presents challenges in debugging and tuning. In complicated systems with various inputs, implementing the lookup table approximation approach can be expensive due to the need for huge memories to store the table.

Fuzzy control, utilizing fuzzy logic, often outperforms linear, piecewise linear, or lookup table approximation approaches in various applications due to its ability to efficiently integrate linguistic fuzzy information provided by human specialists. Intelligent control has been developed to utilize expert knowledge inside the control field. Tools for applying intelligent control include artificial neural networks, genetic algorithms, and fuzzy logic. An artificial neural network is a learning device designed based on the functionality of human brains and their components. The genetic algorithm is based on Charles Robert Darwin's theory of evolution and operates by generating numerous stochastic selection parameters for a given problem. Expert knowledge is typically conveyed using fuzzy terms such as tiny, large, rapid, and similar descriptors. Thus, fuzzy control is better suited for implementing intelligent control using expert information compared to artificial neural networks and evolutionary algorithms.

Fuzzy controllers can execute nonlinear control actions due to the capability of fuzzy logic systems to uniformly approximate any nonlinear function across a compact set with any desired degree of accuracy. If the parameters of a fuzzy controller are selected with care, it is feasible to create a fuzzy controller for nonlinear dynamic systems. Fuzzy controllers are recognized for their robustness against system disturbances due to their operations being guided by fuzzy rules. Fuzzy controllers are easily configurable because to the simplicity of understanding and modifying their principles.

Complex dynamic systems require fuzzy schemes to assess qualitative behaviors using fuzzy algorithms, as proposed by Zadeh [6]. Mamdani and Assilian suggested a fuzzy system with fuzzifier and defuzzifier based on a previous study. They utilized fuzzy logic to regulate a dynamic plant, sparking significant interest among scholars. A fuzzy system is a system that is directly associated with fuzzy ideas such as fuzzy sets, linguistic variables, and fuzzy logic. The fuzzifier function maps precise points to fuzzy sets, while the defuzzifier function maps fuzzy sets to precise points.

Many successful applications of fuzzy control have led to a greater demand for theoretical studies on the stability and performance of fuzzy control systems. Primarily, the stability of fuzzy control systems is typically confirmed through theoretical reasoning. Various important studies have focused on creating stabilizing controllers for fuzzy systems with rigorous stability demonstrations. These studies often utilize the TS-type fuzzy model introduced by Takagi and Sugeno to represent fuzzy systems. The authors developed TS-type fuzzy models for dynamic systems using IF-THEN fuzzy implication and fuzzy inference. They also created stabilizing control rules for these models. They implemented stabilizing control rules designed for TS-type fuzzy models on dynamic systems. Expert knowledge can be utilized in fuzzy control design for dynamic systems through language expressions like the IF-THEN fuzzy implication[5]..

Optimality is a crucial aspect in designing fuzzy control systems, with the typical linear optimum control method being commonly utilized to create the optimal control law for TS-type fuzzy systems. Wang [17] addressed the optimality of fuzzy control systems by creating an optimal fuzzy controller for linear time-invariant systems using the Pontryagin minimum principle. The design strategy described in reference [17] is not practically useful since using a fuzzy controller developed for linear systems directly on nonlinear fuzzy systems may not be effective. Wu and Lin [18] introduced a design method for optimum controllers for continuous- and discrete-time fuzzy systems, drawing from linear quadratic optimal control theory [16]. [18] primarily focuses on finding the best controller that minimizes a certain performance index by solving either the matrix Riccati differential equations or the steady-state algebraic Riccati equations. Wu and Lin [19] studied a quadratic optimal control problem for continuous-time fuzzy systems using a linear-like synthetic matrix form. They developed a design scheme for the optimal fuzzy controller under finite or infinite horizon using the calculus-of-variation method. The research in [19] involves solving a steady-state algebraic Riccati-like equation using an efficient algorithm to develop the best global fuzzy controller. Park et al. [20] studied the optimal control problem for continuous-time TS-type fuzzy systems. The design method in reference [20] has minimal redundancy in selecting feedback gains and necessitates high feedback gains, which are the primary disadvantages in using [20] for creating optimal controllers for fuzzy systems. Kim and Rhee introduced a response surface methodology and used it to create an ideal fuzzy controller for a plant. Wu and Lin [22] suggested a method for creating an optimal discrete-time fuzzy controller to manage and stabilize a nonlinear discrete-time TS-type fuzzy system with finite or infinite horizon time. Chen and Liu [23] researched guaranteed cost control for TS-type fuzzy systems with a time-varying delayed state. Mirzaei and colleagues (2014) introduced an enhanced fuzzy controller for antilock braking systems to enhance vehicle control in emergency braking situations. Lin, Wang, and Lee [25] studied a geometric property of time-optimal control problems in the TS-type fuzzy model using Lie algebra. They determined that the time-optimal controller is of the bang-bang type with a finite number of switching by utilizing the maximum principle. Mostefai et al. [26] introduced a fuzzy observer-based optimum control strategy to address nonlinear friction in a robot joint structure using a fuzzy local modeling technique. Zhu (27) examined a fuzzy optimum control issue using a multistage fuzzy system to maximize the expected value of a fuzzy objective function within the constraints of the system. Esfahani and Sichani [28] researched the optimal fuzzy $H\infty$-tracking control architecture for nonlinear systems represented by the

TS fuzzy modeling scheme. This researchhave significantly advanced the optimum control of fuzzy systems.

TS-type fuzzy systems are inherently nonlinear because of the IF-THEN fuzzy implication and fuzzy inference. Therefore, a nonlinear optimum control method is appropriate for developing an optimal control law for TS-type fuzzy systems. When designing an optimal control law for TS-type fuzzy systems, it is important that the control design enables us to regulate the convergence rates of state trajectories towards an equilibrium point. The decay rate of the closed-loop dynamics can be utilized to meet this need in the control design. The author is inspired by these results to investigate the optimal control of TS-type fuzzy systems that can achieve rapid convergence of state trajectories to an equilibrium point in this chapter.

In this chapter, the author introduces a theoretical finding about the best way to govern nonlinear dynamic systems. The author introduces the optimal control problem for nonlinear dynamic systems and solves it using dynamic programming and the inverse optimal technique. Kalman introduced the inverse optimum approach to determine gain and phase margins of a linear quadratic regulator. The traditional direct optimum method involves finding a stabilizing controller that minimizes a specified performance index. Conversely, the inverse optimum method circumvents solving the Hamilton-Jacobi-Bellman (H-J-B) equation numerically by identifying a stabilizing controller initially and then proving its optimality based on a subsequently discovered performance index. The author uses dynamic programming to obtain the Hamilton-Jacobi-Bellman equation for the optimal control issue in nonlinear dynamic systems in this chapter. The author introduces an analytical method to solve the H-J-B equation using the inverse optimum approach, establishing a systematic process for constructing the optimal controller for nonlinear dynamic systems. The optimal controller obtained is in the form of a state feedback LgV controller with a modified control gain structure.

The author uses the theoretical findings in this chapter to develop an optimal control design for TS-type fuzzy systems. This design ensures global asymptotic stability of an equilibrium point and optimality in relation to a cost function. The cost function includes penalties on state and control input vectors and ensures rapid convergence of state trajectories to the equilibrium point. The issue in this optimum control design for TS-type fuzzy systems is formulated as a problem based on linear matrix inequality (LMI). The ideal controller can be determined using a straightforward controller design process outlined as Linear Matrix Inequalities (LMIs). Control designs that incorporate Linear Matrix Inequalities (LMIs) are highly practical due to the efficient solvability of LMIs using modern interior-point methods. An algorithm from the interior-point approaches is detailed in reference [34]. This algorithm is implemented in the LMI Control Toolbox of MATLAB [35], which will serve as the solver for the LMI-based problem in the optimal control design.

The optimal controller for a nonlinear dynamic system is typically designed based on a linearized design-model, obtained through conventional linear approximation techniques. This approach is used due to the complexity of solving the nonlinear optimal control problem associated with a nonlinear dynamic system. The optimal controller created for a linearized design-model ensures optimality only at the equilibrium point utilized to construct the linearized model. The TS-type fuzzy model is considered a superior design-model for approximating nonlinear dynamic systems compared to conventional linear approximation techniques. It accurately represents the essential features of a nonlinear dynamic system using fuzzy IF-THEN rules, making it valid across various operating points within fuzzy sets. When designing an optimal controller for a nonlinear dynamic system, it is sufficient to approximate the system using a TS-type fuzzy model and utilize this model instead of a linearized design-model. The ideal controller for a TS-type fuzzy system is expected to have a broader range of optimality compared to a controller created for a linearized design-model. This is because the former ensures optimality across various operating points inside fuzzy sets.

This chapter uses the three-axis attitude stabilization problem of a stiff body as a control design example to demonstrate the optimal control design method for TS-type fuzzy systems. The motion of a rigid body's attitude is described by two equations: Euler's dynamic equation, which explains the change in angular velocity vector over time, and the kinematic equation, which connects the rate of change of orientation angles to the angular velocity vector. Various kinematic parameterizations are available for representing the orientation angles of a rigid body, including Euler angles, Gibbs vector, Cayley-Rodrigues parameters, modified Rodrigues parameters, and quaternions (also known as Euler parameters). The former are singular three-dimensional parameter representations, while the latter is a nonsingular four-dimensional parameter representation. Three-dimensional parameter representations can have singular orientations due to the singularity of the Jacobian matrix in certain orientations. Conversely, a quaternion is comprised of four parameters that must adhere to the unit length restriction and serves as a globally unique parameter for defining the body's orientation.

This chapter examines the equations governing the motion of a rigid body, encompassing both dynamics and kinematics. This chapter describes the kinematic equation of a rigid body using quaternions. This chapter discusses the equations of motion of a rigid body in cascade interconnection. The backstepping method from reference [39] can be effectively used to implement the optimal control design method for the three-axis attitude stabilization problem of a rigid body.

Various academics have investigated the ideal attitude stabilization problem of a rigid body [40, 41, 42]. Several studies have examined performance metrics like time and fuel when formulating the optimal attitude stabilization problem of a rigid body. These studies have focused on optimal regulation problems for the angular velocity subsystem of a rigid body and quadratic performance indexes.

Several scholars have studied the optimal attitude control problem of a rigid body's total attitude motion, encompassing both dynamics and kinematics. Carrington and Junkins [50] utilized a polynomial expansion method to approximate the solution of the Hamilton-Jacobi-Bellman equation. Rotea et al. demonstrated that incorporating a logarithmic term into the kinematic parameters of Lyapunov functions leads to linear controllers with a finite quadratic performance index. The generic quadratic performance index also includes conditions that ensure the presence of a linear and suboptimal stabilizing controller. Tsiotras developed a new set of feedback control laws that globally stabilize and a group of optimal control laws that exponentially stabilize the full attitude motion of a nonsymmetric rigid body. Tsiotras [53] later proposed a partial solution to the optimal regulation problem of a spinning rigid body by utilizing the natural decomposition of the full attitude motion into its kinematics and dynamics systems, along with the intrinsic passivity features of these systems. Bharadwaj et al. [54] introduced novel attitude control rules that provide global stabilization using minimum and exponential coordinates. Park and Tahk (55) addressed the issue of stabilizing the attitude of a rigid body along three axes in the presence of unpredictable inertia. They introduced a set of novel attitude control laws with more flexible feedback gain configurations. Park and Tahk expanded their robust attitude control method to an optimum attitude control scheme by applying the Hamilton-Jacobi theory. Park et al. [58] were the first to introduce a game-theoretic method for achieving resilient and optimal attitude stabilization of a rigid body in the presence of external disturbances.

In robot arm control, the arms and hand fingers operate as actuators to manipulate the attitude of the held object. Therefore, findings on the attitude control of a rigid body can be extended to the attitude control of a rigid payload held by the robot arm. Many studies have been conducted on the attitude control problem of a rigid body in relation to this topic, with notable references including [60, 61].

The remainder of this chapter is structured as follows. The author introduces a theoretical finding about the optimal control of nonlinear dynamic systems in Section 2. Section 3 describes TS-type fuzzy systems and discusses an optimal control design for these systems. In Section 4, the author examines the three-axis attitude stabilization problem of a stiff body as a control design example and demonstrates the efficiency of optimal control design for TS-type fuzzy systems. In Section 5, the author ends this chapter with concluding observations.

2. NONLINEAR OPTIMAL CONTROL

Consider the nonlinear dynamic system given by

$$\dot{x}(t) = f(x(t)) + g(x(t))u(t), \tag{1}$$

where f:Rn Rn and g:Rn Rn×p are smooth, vector- and matrix-valued functions, respectively, and f(0)=0. Moreover, x(t)∈Rn and u(t)∈Rp are the state and control input vectors, respectively. Throughout this chapter, we use the definitions of

$$L_f V(x(t)) \triangleq \left(\frac{\partial V(x(t))}{\partial x(t)}\right) f(x(t)) \text{ and}$$

$$L_g V(x(t)) \triangleq \left(\frac{\partial V(x(t))}{\partial x(t)}\right) g(x(t)), \text{ where}$$

V:Rn R is a scalar function [4].

In general, we can find the optimal control law for the nonlinear dynamic system in (1) by numerically solving the corresponding Hamilton-Jacobi-Bellman (H-J-B) equation. However, this is a difficult task, and, thus, we may need a simple and efficient method to find optimal control law for the nonlinear dynamic system in (1). In the following, the author presents a theory to provide the optimal control law for the nonlinear dynamic system in (1) by circumventing the task of numerically solving the H-J-B equation.

Proposition 1 [62]: For the nonlinear dynamic system in (1), suppose that there exists a radially unbounded and positive definite function V(x(t)) that has continuous, first, partial derivatives with respect to x(t) and the feedback control u(t)=- R-1[LgV(x(t))]T, where ≥1 is a constant and R=RT>0 is a positive definite matrix, achieves global asymptotic stability of the equilibrium point x(t)=0 for the system in (1) such that:

$$\dot{V}(x(t))\Big|_{u(t)=-\alpha R^{-1}[L_g V(x(t))]^T} = L_f V(x(t)) - \alpha[L_g V(x(t))]R^{-1}[L_g V(x(t))]^T < 0 \tag{2}$$

for all x(t)≠0. Then, the control law

$$u^*(t) = -2\alpha R^{-1}[L_g V(x(t))]^T \tag{3}$$

is the optimal, globally asymptotically stabilizing control law for the system in (1) that minimizes the cost function

$$P = \int_0^\infty \left[l(x(t)) + u(t)^T R u(t)\right] dt, \tag{4}$$

where l(x(t)) is given by

$$l(x(t)) = -4\alpha^2\left(L_fV(x(t)) - \alpha[L_gV(x(t))]R^{-1}[L_gV(x(t))]^T\right)$$
$$+ 4\alpha^2(\alpha-1)[L_gV(x(t))]R^{-1}[L_gV(x(t))]^T \tag{5}$$
$$> 0$$

for all x(t)≠0 and ≥1 .

Proof: First, the following condition holds by (2):

$$\dot{V}(x(t))|_{u(t)=u^*(t)} = \left(L_fV(x(t)) + \frac{1}{2}[L_gV(x(t))]u^*(t)\right) + \frac{1}{2}[L_gV(x(t))]u^*(t)$$
$$= \left(\dot{V}(x(t))\Big|_{u(t)=-\alpha R^{-1}[L_gV(x(t))]}\right) - \alpha[L_gV(x(t))]R^{-1}[L_gV(x(t))]^T$$
$$< -\alpha[L_gV(x(t))]R^{-1}[L_gV(x(t))]^T < 0 \tag{6}$$

for all x t ≠ 0 and ≥ 1 . Since V x t is a radially unbounded and positive definite function, the condition in (6) guarantees that the control law u ∗ t in (3) is a globally asymptotically stabilizing control law for the system in (1) by the Lyapunov's stability theorem [4].

Next, define $W(x(t)) \triangleq 4\alpha^2V(x(t))$ and consider the following H-J-B equation associated with the optimal control problem for the system in (1):

$$\min_{u(t)}\left(l(x(t)) + u(t)^TRu(t) + \{L_fW(x(t)) + [L_gW(x(t))]u(t)\}\right) = 0, \quad W(0) = 0. \tag{7}$$

Substituting u∗t in (3) and l x t in (5) into the H-J-B equation in (7) yields

$$\min_{u(t)=u^*(t)}\left(l(x(t)) + u(t)^TRu(t) + \{L_fW(x(t)) + [L_gW(x(t))]u(t)\}\right)$$
$$= \min_{u(t)=u^*(t)}\left(\left\{-L_fW(x(t)) + \frac{1}{4\alpha}[L_gW(x(t))]R^{-1}[L_gW(x(t))]^T\right.\right.$$
$$\left. + \frac{(\alpha-1)}{4\alpha^2}[L_gW(x(t))]R^{-1}[L_gW(x(t))]^T\right\} + \frac{1}{4\alpha^2}[L_gW(x(t))]R^{-1}[L_gW(x(t))]^T$$
$$\left. + \left\{L_fW(x(t)) - \frac{1}{2\alpha}[L_gW(x(t))]R^{-1}[L_gW(x(t))]^T\right\}\right) = 0, \quad W(0) = 0, \tag{8}$$

which implies that u∗t in (3) and in (5) are solutions of the H-J-B equation in (7). In addition, by (2) and the property of R = RT > 0 , l x t in (5) satisfies l(x(t)) > 0 for all x(t) ≠ 0 and ≥ 1 . This completes the proof.

In proposition 1, we see that the globally asymptotically stabilizing control law in (3) for the nonlinear dynamic system in (1) can be found without the task of numerically solving the H-J-B equation in (7) and the control law in (3) is optimal with respect to the cost function in (4). The key point of this work is that we posteriorly determine the penalty on the state vector, which is l(x(t)), rather than we priorly choose it. Sepulcher, Janković, and Kokotović [30] proposed this approach which is referred to as the inverse optimal approach.

It is remarkable that as shown in (4, 5), we can adjust the penalty on the control input vector, which is R , and the penalty on the state vector, which is l(x(t)), with the weight matrix R . Indeed, we can decrease the penalty on the control input vector and increase the penalty on the state vector with a weight matrix R having small values. In this condition, we can obtain a cheap optimal control law requiring a large control effort, and this cheap optimal control law makes the nonlinear dynamic system in (1) stable within a short period of time. Note that the term "cheap" refers to the fact that the control effort is viewed as being cheap. On the other hand, we can increase the penalty on the control input vector and decrease the penalty on the state vector with a weight matrix R having large values. In this condition, we can obtain an expensive optimal control law requiring a small control effort, and this expensive optimal control law makes the nonlinear dynamic system in (1) stable within a long period of time. Note that the term "expensive" refers to the fact that the control effort is viewed as being expensive.

As shown in (3), since the constant α of the optimal control law u ∗ t in (3) plays the role of a feedback gain for u ∗ t even though the weight matrix R is predetermined to impose the penalties on the control input and state vectors, it is also remarkable that the optimal control law u ∗ t in (3) has a relaxed feedback gain structure.

Now, if we consider a practical application, we know that any control law for dynamic systems must provide good convergence rates of state trajectories to an equilibrium point. For achieving this requirement in the design of a control law, the decay rate can be used as a design factor to dominate convergence rates of state trajectories to an equilibrium point. Note that the decay rate of a system is defined to be the largest constant γ > 0 such that $\lim_{t\to\infty} e^{\gamma t}\|x(t)\|_2 = 0$ holds for all trajectories of a system, where $\|x(t)\|_2$denotes the Euclidean norm of x(t). From the definition of decay rate, the convergence rate of the system trajectory to the equilibrium point can be controlled. Also, note that stability of dynamic systems corresponds to a positive decay rate. In the following, the author presents a theory about the decay rate.

Proposition 2 [32]: If there exist a positive definite function V x t and a constant > 0 such that

$$\dot{V}(x(t)) < -2\beta V(x(t)) \tag{9}$$

for all trajectories of a system, then the decay rate of a system is at least .

Proof: If there exist a positive definite function V(x(t)) and a constant >0 such that $\dot{V}(x(t)) < -2\beta V(x(t))$ for all trajectories of a system, then we obtain $V(x(t)) < V(x(0))e^{-2\beta t}$. With a positive definite function $V(x(t)) \triangleq x(t)^T X^{-1} x(t)$, where X = XT > 0 is a positive definite matrix, $V(x(t)) < V(x(0))e^{-2\beta t}$ can be represented by $\left\|X^{-\frac{1}{2}}x(t)\right\|_2^2 < \left\|X^{-\frac{1}{2}}e^{-\beta t}x(0)\right\|_2^2$. Then, by Rayleigh-Ritz theorem [63], we can derive the following:

$$\|x(t)\|_2^2 \leq \frac{\left\|X^{-\frac{1}{2}}x(t)\right\|_2^2}{\lambda_{\min}(X^{-1})} < \frac{\left\|X^{-\frac{1}{2}}e^{-\beta t}x(0)\right\|_2^2}{\lambda_{\min}(X^{-1})} \leq \frac{\left\|X^{-\frac{1}{2}}\right\|_2^2}{\lambda_{\min}(X^{-1})}\left\|e^{-\beta t}x(0)\right\|_2^2 = \frac{\lambda_{\max}(X^{-1})}{\lambda_{\min}(X^{-1})}\|x(0)\|_2^2 e^{-2\beta t},$$

where λmin X−1 and λmax X−1 denote the minimum and maximum eigenvalues of X−1 , respectively. Thus, we obtain $\|x(t)\|_2 < \sqrt{(\lambda_{\max}(X^{-1})/\lambda_{\min}(X^{-1}))}$, and therefore the decay rate of the system is at least . This completes the proof.

3. OPTIMAL CONTROL OF FUZZY SYSTEMS

3.1. Fuzzy Systems

The TS-type fuzzy model and the TS-type fuzzy control law for a system are given by the following IF-THEN fuzzy implications, respectively [15]:

Plant rule i for a system:

IF x1(t) is Mi1 and ··· and xn(t) is Min, THEN

$$\dot{x}(t) = A_i x(t) + B_i u(t),\ i = 1, \cdots, r. \tag{10}$$

◉ *Control law rule i for a system:*

IF x1(t) is Mi1 and ··· and xn(t) is Min, THEN

$$u(t) = -K_i x(t),\ i = 1, \cdots, r. \tag{11}$$

In (10, 11), xi(t) , i = 1 , ··· , n and Mij , i = 1 , ··· , r , j = 1 , ··· , n are state variables and fuzzy sets, respectively, and r is the number of IF-THEN rules. Moreover, A i ∈ Rn×n, B i ∈ Rn×p, and K i ∈ Rp×n.

If we follow the usual fuzzy inference method, we can represent the state equations of the TS-type fuzzy model and the TS-type fuzzy control law for a system as follows, respectively [9]:

$$\dot{x}(t) = \sum_{i=1}^{r} h_i(x(t))A_i x(t) + \sum_{i=1}^{r} h_i(x(t))B_i u(t) \tag{12}$$

and

$$u(t) = -\sum_{i=1}^{r} h_i(x(t))K_i x(t), \tag{13}$$

where hi, i = 1 , ⋯ , r are the normalized weight functions given by

$$h_i(x(t)) \triangleq \frac{\prod_{j=1}^{n} M_{ij}(x_j(t))}{\sum_{i=1}^{r} M_{ij}(x_j(t))}, \quad i = 1, \cdots, r \tag{14}$$

and Mij (xj(t)) denotes the grade of membership of xj(t) in the fuzzy set Mij . Here, hi , i = 1 , ⋯ , r in (14) satisfy hi(x(t)) ≥ 0 , i = 1 , ⋯ , r and $\sum_{i=1}^{r} h_i(x(t)) = 1$ for all x t ∈ Rn.

3.2. Problem Definition

The author defines two kinds of problems considered in this section. The first problem is to design a control law for the TS-type fuzzy system in (12) that achieves the global asymptotic stability of the equilibrium point x t = 0 and minimizes the cost function

$$P = \int_0^{\infty} \left[l(x(t)) + u(t)^T R u(t) \right] dt,$$

where l(x(t)) > 0 for all (x(t)) ≠ 0 and R = RT > 0 is a positive definite matrix. The second problem is that the decay rate of the closed-loop dynamics for the TS-type fuzzy system in (12) should be at least , where > 0 is a constant.

3.3. Optimal Control Design

We can use the results of propositions 1 and 2 to solve the problems defined in Section 3.2. Specifically, we define $f(x(t)) \triangleq \sum_{i=1}^{r} h_i(x(t)) A_i x(t)$ and $g(x(t)) \triangleq \sum_{i=1}^{r} h_i(x(t)) B_i$ for the TS-type fuzzy system in (12). Then, if there exists a radially unbounded and positive definite function V(x(t)) such that the conditions in (2) and (9) hold, then an optimal control law in the form of (3) for the TS-type fuzzy system in (12) can be found. Thus, we have to construct V x t whose time derivative satisfies the conditions in (2, 9). The author uses a quadratic Lyapunov function V(x(t)) = x(t)T X−1 x(t), where X = XT > 0 is a positive definite matrix, as the candidate of such a V(x(t)), and presents the following theory:

Theorem 1 [62]: For the TS-type fuzzy system in (12), suppose that there exists a function V(x(t)) = x(t)T X−1 x(t) with X ∈ Rn×n such that

$$X = X^T > 0, \tag{15}$$

$$\begin{aligned} &A_i X + X A_i^T - 4\alpha B_i R^{-1} B_i^T < 0, \quad i = 1, \cdots, r, \\ &\frac{1}{2}\left(A_i X + X A_i^T + A_j X + X A_j^T\right) - 2\alpha\left(B_i R^{-1} B_j^T + B_j R^{-1} B_i^T\right) < 0, \quad 1 \le i < j \le r, \end{aligned} \tag{16}$$

and

$$\begin{aligned} &A_i X + X A_i^T - 8\alpha B_i R^{-1} B_i^T + 2\beta X < 0, \quad i = 1, \cdots, r, \\ &\frac{1}{2}\left(A_i X + X A_i^T + A_j X + X A_j^T\right) - 4\alpha\left(B_i R^{-1} B_j^T + B_j R^{-1} B_i^T\right) + 2\beta X < 0, \quad 1 \le i < j \le r, \end{aligned} \tag{17}$$

where R = RT > 0 is a positive definite matrix and ≥ 1 and > 0 are constants. Then the control law

$$u^*(t) = -\sum_{i=1}^{r} h_i(x(t))K_i x(t), \tag{18}$$

where $K_i \triangleq 4\alpha R^{-1}B_i^T X^{-1}, i = 1, \cdots, r$ is the optimal, globally asymptotically stabilizing control law for the TS-type fuzzy system in (12) that minimizes the cost function

$$P = \int_0^\infty \left[l(x(t)) + u(t)^T R u(t)\right] dt, \tag{19}$$

where l(x(t)) is given by

$$\begin{aligned} l(x(t)) &= -4\alpha^2 x(t)^T \left\{ \sum_{i=1}^{r} h_i^2(x(t))(G_{ii}^T X^{-1} + X^{-1}G_{ii}) \right. \\ &\quad \left. + 2\sum_{i<j}^{r} h_i(x(t))h_j(x(t)) \left[\left(\frac{G_{ij}+G_{ji}}{2}\right)^T X^{-1} + X^{-1}\left(\frac{G_{ij}+G_{ji}}{2}\right) \right] \right\} x(t) \\ &\quad + 4\alpha^2(\alpha-1)x(t)^T \left\{ 4\sum_{i=1}^{r}\sum_{j=1}^{r} h_i(x(t))h_j(x(t)) \left(X^{-1}B_i R^{-1} B_j^T X^{-1}\right) \right\} x(t) \\ &> 0 \end{aligned} \tag{20}$$

for all x(t) ≠ 0 and ≥ 1 , where $G_{ij} \triangleq A_i - 2\alpha B_i R^{-1} B_j^T X^{-1}$, and the decay rate of the closed-loop dynamics for the TS-type fuzzy system in (12) with the control law u ∗ t in (18) is at least , where > 0 is a constant.

Proof: Suppose that $V(x(t)) = x(t)^T X^{-1} x(t)$, where X = XT > 0 is a positive definite matrix. Then, from proposition 1, the control law u ∗ t in (3) with $g(x(t)) \triangleq \sum_{i=1}^{r} h_i(x(t))B_i$ and $V(x(t)) = x(t)^T X^{-1} x(t)$ becomes

$$\begin{aligned} u^*(t) &= -2\alpha R^{-1}\left[L_g V(x(t))\right]^T \\ &= -2\alpha R^{-1}\left[2x(t)^T X^{-1}\left(\sum_{i=1}^{r} h_i(x(t))B_i\right)\right]^T \\ &= -\sum_{i=1}^{r} h_i(x(t))(4\alpha R^{-1}B_i^T X^{-1})x(t) \\ &\triangleq -\sum_{i=1}^{r} h_i(x(t))K_i x(t), \end{aligned} \tag{21}$$

where $K_i \triangleq 4\alpha R^{-1}B_i^T X^{-1}$, $i = 1, \cdots, r$ and ≥ 1 is a constant.

Now, from propositicns 1 and 2, assume that there exists a positive definite matrix X = XT > 0 such that

$$\begin{aligned}\dot{V}(x(t))\Big|_{u(t)=\frac{1}{2}u^*(t)} &= L_f V(x(t)) + \frac{1}{2}\left[L_g V(x(t))\right]u^*(t) \\ &= x(t)^T\Big\{\sum_{i=1}^{r} h_i(x(t))^2 (G_{ii}^T X^{-1} + X^{-1}G_{ii}) \\ &+ 2\sum_{i<j}^{r} h_i(x(t))h_j(x(t))\left[\left(\frac{G_{ij}+G_{ji}}{2}\right)^T X^{-1} + X^{-1}\left(\frac{G_{ij}+G_{ji}}{2}\right)\right]\Big\}x(t) \\ &< 0\end{aligned} \tag{22}$$

for all x(t) ≠ 0 , where $G_{ij} \triangleq A_i - \frac{1}{2}B_i\left(4\alpha R^{-1}B_j^T X^{-1}\right) = A_i - \frac{1}{2}B_iK_j$, and

$$\begin{aligned}\dot{V}(x(t))\big|_{u(t)=u^*(t)} &= L_f V(x(t)) + \left[L_g V(x(t))\right]u^*(t) \\ &= x(t)^T\Big\{\sum_{i=1}^{r} h_i^2(x(t))(Q_{ii}^T X^{-1} + X^{-1}Q_{ii}) \\ &+ 2\sum_{i<j}^{r} h_i(x(t))h_j(x(t))\left[\left(\frac{Q_{ij}+Q_{ji}}{2}\right)^T X^{-1} + X^{-1}\left(\frac{Q_{ij}+Q_{ji}}{2}\right)\right]\Big\}x(t) \\ &< -2\beta x(t)^T X^{-1}x(t)\end{aligned} \tag{23}$$

for all $x(t)$, where $Q_{ij} \triangleq A_i - B_i\left(4\alpha R^{-1}B_j^T X^{-1}\right) = A_i - B_iK_j$ and $\beta > 0$ is a constant. Since the normalized weight functions h_i, $i = 1, \cdots, r$ in (22) and (23) satisfy $h_i(x(t))h_j(x(t)) \geq 0, i = 1, \cdots, r, j = 1, \cdots, r$ and $\sum_{i=1}^{r}\sum_{j=1}^{r} h_i(x(t))h_j(x(t)) = 1$ (i.e., $\sum_{i=1}^{r}\sum_{j=1}^{r} h_i(x(t))h_j(x(t)) = \sum_{i=1}^{r} h_i^2(x(t)) + 2\sum_{i<j}^{r} h_i(x(t))h_j(x(t)) = 1$) for all $x(t) \in R^n$, sufficient conditions for satisfying (22) and (23) are

∈ Rn , sufficient conditions for satisfying (22) and (23) are

$$\begin{aligned}&G_{ii}^T X^{-1} + X^{-1}G_{ii} < 0, \quad i = 1, \cdots, r, \\ &\left(\frac{G_{ij}+G_{ji}}{2}\right)^T X^{-1} + X^{-1}\left(\frac{G_{ij}+G_{ji}}{2}\right) < 0, \quad 1 \leq i < j \leq r\end{aligned} \tag{24}$$

and

$$\begin{aligned}&Q_{ii}^T X^{-1} + X^{-1}Q_{ii} + 2\beta X^{-1} < 0, \quad i = 1, \cdots, r, \\ &\left(\frac{Q_{ij}+Q_{ji}}{2}\right)^T X^{-1} + X^{-1}\left(\frac{Q_{ij}+Q_{ji}}{2}\right) + 2\beta X^{-1} < 0, \quad 1 \leq i < j \leq r,\end{aligned} \tag{25}$$

respectively. If all sets of inequalities in (24, 25) are pre- and post-multiplied by a positive definite matrix X = XT > 0 , then we can transform the nonlinear conditions in (24, 25) into the linear conditions in (16, 17), respectively. Thus, by the results of propositions 1 and 2, we see that the control law u ∗ t in (21) with a positive definite matrix X = XT > 0 satisfying the conditions in (16, 17) becomes the optimal, globally asymptotically stabilizing control law for the TS-type fuzzy system in (12) that minimizes the cost function in (19), where l(x(t)) in (20) comes from (5) and satisfies l(x(t)) > 0 for all x(t) ≠ 0 and ≥ 1 by (22) and the property of R = RT > 0 , and the decay rate of the closed-loop dynamics for the TS-type fuzzy system in (12) with the control law u ∗ t in (21) becomes at least > 0 . This completes the proof.

Now, consider the TS-type fuzzy system with a common input matrix, which is described by

$$\dot{x}(t) = \sum_{i=1}^{r} h_i(x(t))A_i x(t) + Bu(t). \tag{26}$$

Then, the author presents the following theory:

Theorem 2 [62]: For the TS-type fuzzy system in (26), suppose that there exists a function V(x(t)) = x(t)T X−1 x(t) with X ∈ Rn×n such that

$$X = X^T > 0, \tag{27}$$

$$A_i X + X A_i^T - 4\alpha B R^{-1} B^T < 0, \quad i = 1, \cdots, r, \tag{28}$$

and

$$A_i X + X A_i^T - 8\alpha B R^{-1} B^T + 2\beta X < 0, \quad i = 1, \cdots, r, \tag{29}$$

where R = RT > 0 is a positive definite matrix and ≥ 1 and > 0 are constants. Then the control law:

$$u^*(t) = -Kx(t), \tag{30}$$

where $K \triangleq 4\alpha R^{-1} B^T X^{-1}$ is the optimal, globally asymptotically stabilizing control law for the TS-type fuzzy system in (26) that minimizes the cost function

$$P = \int_0^\infty \left[l(x(t)) + u(t)^T R u(t) \right] dt, \tag{31}$$

where l(x(t)) is given by

$$\begin{aligned} l(x(t)) &= -4\alpha^2 x(t)^T \left\{ \sum_{i=1}^{r} h_i(x(t)) (A_i^T X^{-1} + X^{-1} A_i) - 4\alpha X^{-1} B R^{-1} B^T X^{-1} \right\} x(t) \\ &+ 4\alpha^2 (\alpha - 1) x(t)^T (4 X^{-1} B R^{-1} B^T X^{-1}) x(t) \\ &> 0 \end{aligned} \tag{32}$$

for all x(t) ≠ 0 and ≥ 1 , and the decay rate of the closed-loop dynamics for the TS-type fuzzy system in (26) with the control law u ∗ t in (30) is at least , where > 0 is a constant.

Proof: Suppose that V(x(t)) = x(t)T X−1 x(t), where X = XT > 0 is a positive definite matrix. Then, from proposition 1, the control law u ∗ t in (3) with $g(x(t)) \triangleq B$ and $V(x(t)) = x(t)^T X^{-1} x(t)$ becomes

$$
\begin{aligned}
u^*(t) &= -2\alpha R^{-1}\left[L_g V(x(t))\right]^T \\
&= -2\alpha R^{-1}\left[2x(t)^T X^{-1} B\right]^T \\
&= -4\alpha R^{-1} B^T X^{-1} x(t) \\
&\triangleq -Kx(t),
\end{aligned}
\tag{33}
$$

where $K \triangleq 4\alpha R^{-1} B^T X^{-1}$ and ≥ 1 is a constant.

Now, from propositions 1 and 2, assume that there exists a positive definite matrix X = XT > 0 such that

$$
\begin{aligned}
\dot{V}(x(t))\big|_{u(t)=\frac{1}{2}u^*(t)} &= L_f V(x(t)) + \frac{1}{2}\left[L_g V(x(t))\right]u^*(t) \\
&= x(t)^T \left\{ \sum_{i=1}^{r} h_i(x(t))\left(A_i^T X^{-1} + X^{-1} A_i\right) - 4\alpha X^{-1} B R^{-1} B^T X^{-1} \right\} x(t) \\
&< 0
\end{aligned}
\tag{34}
$$

for all x(t) ≠ 0 and

$$
\begin{aligned}
\dot{V}(x(t))\big|_{u(t)=u^*(t)} &= L_f V(x(t)) + \left[L_g V(x(t))\right]u^*(t) \\
&= x(t)^T \left\{ \sum_{i=1}^{r} h_i(x(t))\left(A_i^T X^{-1} + X^{-1} A_i\right) - 8\alpha X^{-1} B R^{-1} B^T X^{-1} \right\} x(t) \\
&< -2\beta x(t)^T X^{-1} x(t)
\end{aligned}
\tag{35}
$$

for all x(t), where > 0 is a constant. Since the normalized weight functions hi, i = 1 , ⋯ , r in (34, 35) satisfy hi(x(t)) ≥ 0 , i = 1 , ⋯ , r and $\sum_{i=1}^{r} h_i(x(t)) = 1$ for all x t ∈ Rn, sufficient conditions for satisfying (34, 35) are

$$A_i^T X^{-1} + X^{-1} A_i - 4\alpha X^{-1} B R^{-1} B^T X^{-1} < 0, \quad i = 1, \cdots, r \tag{36}$$

and

$$A_i^T X^{-1} + X^{-1} A_i - 8\alpha X^{-1} B R^{-1} B^T X^{-1} + 2\beta X^{-1} < 0, \quad i = 1, \cdots, r, \tag{37}$$

respectively. If all sets of inequalities in (36, 37) are pre- and post-multiplied by a positive definite matrix X = XT > 0 , then we can transform the nonlinear conditions in (36, 37) into the linear conditions in (28, 29), respectively. Therefore, by the results of propositions 1 and 2, we see that the control law u ∗ t in (33) with a positive definite matrix X = XT > 0 satisfying the conditions in (28, 29) becomes the optimal, globally asymptotically stabilizing control law for the TS-type fuzzy system in (26) that minimizes the cost function in (31), where l(x(t)) in (32) comes from (5) and satisfies l(x(t)) > 0 for all x(t) ≠ 0 and ≥ 1 by (34) and the property of R = RT > 0 , and the decay rate of the closed-loop dynamics for the TS-type fuzzy system in (26) with the control law u ∗ t in (33) becomes at least > 0 . This completes the proof.

Note that the problem appearing in Theorems 1 and 2 is to find a matrix X ∈ Rn×n subject to some linear constraints in the form of linear matrix inequality (LMI). Therefore, this problem is an LMI-based problem [32], and we can efficiently solve the LMI-based problem by the LMI Control Toolbox of MATLAB [35]. In this chapter, the author uses the LMI Control Toolbox of MATLAB [35] as the solver for the LMI-based problem in the optimal control design.

4. A CONTROL DESIGN EXAMPLE

The author examines the three-axis attitude stabilization problem of a stiff body as a control design example and demonstrates the efficacy of the optimal control design for TS-type fuzzy systems discussed in Section 3.

4.1. Rigid Body Model

First, the dynamic equation of the rotational motion of a rigid body is described as follows [36]:

$$\dot{\omega}(t) = J^{-1}\Omega(\omega(t))J\omega(t) + J^{-1}u(t), \tag{38}$$

where ω(t) = [ω1(t) ω2(t) ω3(t)]T ∈ R3 is the angular velocity vector of the body in the body-fixed frame, u(t) = [u1(t) u2(t) u3(t)]T ∈ R3 is the control torque vector of the body, and J ∈ R3×3 is the inertia matrix of the body and satisfies J = JT > 0 . And Ω(ω(t)) ∈ R3×3 denotes a skew-symmetric matrix defined by

$$\Omega(\omega(t)) \triangleq \begin{bmatrix} 0 & \omega_3(t) & -\omega_2(t) \\ -\omega_3(t) & 0 & \omega_1(t) \\ \omega_2(t) & -\omega_1(t) & 0 \end{bmatrix}$$

and has the property of

$$\Omega(\omega(t))^T\omega(t) \equiv 0, \quad \forall\omega(t) \in R^3. \tag{39}$$

Second, the kinematic equation of rotational motion of a rigid body described in terms of the quaternion is given as follows [36]:

$$\dot{q}(t) = \frac{1}{2}F(q(t))\omega(t), \tag{40}$$

where $q(t) = [q_1(t)\, q_2(t)\, q_3(t)\, q_4(t)]^T \triangleq \left[q_1(t)\, q_v(t)^T\right]^T \in R^4$ is the quaternion and F(q(t)) : R4 R4×3 denotes the kinematics Jacobian matrix defined as

$$F(q(t)) \triangleq \begin{bmatrix} -q_v(t)^T \\ q_1(t) I_{3\times 3} + \Omega(q_v(t))^T \end{bmatrix}, \tag{41}$$

where I3×3 denotes the 3 × 3 identity matrix.

With the notations of the Euler axis $\hat{e} \in R^3$ and Euler angle ϕ ∈ R , we define the quaternion by $q_1(t) \triangleq \cos(\phi/2)$ and $q_v(t) \triangleq \hat{e} \sin(\phi/2)$. The quaternion q(t) is subject to the unit length constraint of $\|q(t)\|_2$ for all t ≥ 0 and is a kinematic parameter set that can represent the orientation of a body and [36]. From the definition of the quaternion, we see that q1(t) satisfies q1(t) ≥ 0 for all - π rad ≤ ϕ ≤ π rad , which describes all eigenaxis rotations [36]. Thus, we can write $q_1(t) = \sqrt{1 - \|q_v(t)\|_2^2}$ for all - π rad ≤ ϕ ≤ π rad . .

4.2. Optimal Control Design

First, it is observed that two state equations given by (38) and (41) represent a system in cascade interconnection. That is, the angular velocity vector indirectly controls the kinematics system in (41). Thus, the angular velocity vector can be regarded as a virtual control input of the kinematics system in (41) to stabilize the kinematics system in (41). This observation gives the following theorem: Theorem 3 [62]: Consider the kinematics system in (41) with ω t to be the control input, and let the control law for the kinematics system in (41) be

$$\omega_v(t) = -k_1 q_v(t), \tag{42}$$

where k 1 > 0 is a constant. Then, ωv(t) in (43) is the global asymptotic stabilizing control law for the kinematics system in (41).

Proof: With the control law ωv(t) in (43), the closed-loop system of the kinematics system in (41) becomes

$$\dot{q}(t) = -\frac{1}{2} k_1 F(q(t)) q_v(t) = -\frac{1}{2} k_1 \begin{bmatrix} -q_v(t) \\ q_1(t) I_{3\times 3} + \Omega(q_v(t))^T \end{bmatrix} q_v(t). \tag{43}$$

Now, consider the Lyapunov function candidate V(q(t)) = (q1(t) - 12) + $\|q_v(t)\|_2^2$. Taking the time derivative of V q t along a nonzero trajectory of the closed-loop system in (44) and using the property of $\Omega(q_v(t)) q_v(t) \equiv 0$ for all $q_v(t) \in R^3$, then the following condition holds:

$$\begin{aligned} \dot{V}(q(t)) &= 2(q_1(t) - 1)\dot{q}_1(t) + 2 q_v(t)^T \dot{q}_v(t) \\ &= 2 q_1(t) \left(\frac{1}{2} k_1 q_v(t)^T q_v(t)\right) - 2\left(\frac{1}{2} k_1 q_v(t)^T q_v(t)\right) \end{aligned}$$

$$+ 2q_v(t)^T\left(-\frac{1}{2}k_1q_1(t)q_v(t) + \frac{1}{2}k_1\Omega(q_v(t))q_v(t)\right)$$

$$= -k_1q_v(t)^Tq_v(t) < 0$$

for all qv(t) ≠ 0 and all k1 > 0 . Then, global asymptotic stability of the closed-loop dynamics in (44) follows from the Barbashin-Krasovskii theorem [4]. This completes the proof.

Next, we have to stabilize the dynamics system in (38) with making ω(t) in (38) follow ωv(t) in (43), and this is a backstepping problem [39]. For solving this problem, the author defines the new variable δ t as $\delta(t) \triangleq \omega(t) - \omega_v(t) = \omega(t) + k_1q_v(t)$. For convenience of notation, the author defines $x_1(t) \triangleq \delta_1(t)$, $x_2(t) \triangleq \delta_2(t)$, $x_3(t) \triangleq \delta_3(t)$, $x_4(t) \triangleq q_2(t)$, $x_5(t) \triangleq q_3(t)$, $x_6(t) \triangleq q_4(t)$, $x_\delta(t) \triangleq [x_1(t) x_2(t)\, x_3(t)]^T$, and $x_{q_v}(t) \triangleq [x_4(t)\, x_5(t)\, x_6(t)]^T$. Then, with $x(t) \triangleq \left[x_\delta(t)^T\, x_{q_v}(t)^T\right]^T$ and $q_1(t) = \sqrt{1 - \|x_{q_v}(t)\|_2^2}$ for all - π rad ≤ ϕ ≤ π rad , the author represents the state equation for x t by

$$\dot{x}(t) = A(x(t))x(t) + Bu(t), \tag{44}$$

where A(x(t)) ∈ R6×6 and B ∈ R6×3 are defined in (46) and (47), respectively. In (47), 03×3 denotes the 3 × 3 zero matrix:

$$A(x(t)) \triangleq \begin{bmatrix} \begin{Bmatrix} J^{-1}\Omega(x_\delta(t) - k_1x_{q_v}(t))J \\ +\frac{1}{2}k_1\Omega(x_{q_v}(t))^T \\ +\left(\frac{1}{2}k_1\sqrt{1 - \|x_{q_v}(t)\|_2^2}\right)I_{3\times3} \end{Bmatrix}, & \begin{Bmatrix} -k_1J^{-1}\Omega(x_\delta(t) - k_1x_{q_v}(t))J \\ -\frac{1}{2}k_1^2\Omega(x_{q_v}(t))^T \\ -\left(\frac{1}{2}k_1^2\sqrt{1 - \|x_{q_v}(t)\|_2^2}\right)I_{3\times3} \end{Bmatrix} \\ \begin{Bmatrix} \frac{1}{2}\Omega(x_{q_v}(t))^T \\ +\left(\frac{1}{2}k_1\sqrt{1 - \|x_{q_v}(t)\|_2^2}\right)I_{3\times3} \end{Bmatrix}, & \begin{Bmatrix} -\frac{1}{2}k_1\Omega(x_{q_v}(t))^T \\ -\left(\frac{1}{2}k_1\sqrt{1 - \|x_{q_v}(t)\|_2^2}\right)I_{3\times3} \end{Bmatrix} \end{bmatrix}. \tag{45}$$

$$B \triangleq \begin{bmatrix} J^{-1} \\ 0_{3\times3} \end{bmatrix}. \tag{46}$$

As a numerical example, the author assumes that

$$J = \mathrm{diag}[10, 15, 20]\ \mathrm{kg}\cdot\mathrm{m}^2, \tag{47}$$

where diag means the diagonal matrix, k1 = 0.2, and xδi , xqvi ∈[-0.5, 0.5], i = 1 , ⋯ , 3. If we do sampling A(x(t)) in (46) at nine operating points of $\left[x_{\delta_i}, x_{q_{v_i}}\right] = [0\ 0], [0\ 0.5], [0\ -0.5], [0.25\ 0], [0.25\ 0.5], [-0.25\ 0], [-0.25\ -0.5], [0.5\ 0.5], [-0.5\ -0.5], i = 1, \cdots, 3$ with the given k1 = 0.2 and J in (48), we can obtain the following TS-type fuzzy model for the system in (38):

Plant rule i for the system in (45):

IF x1(t) is Mi1 and ⋯ and x6(t) is Mi6, THEN

$$\dot{x}(t) = A_i x(t) + Bu(t), i = 1, \cdots, 9. \tag{48}$$

In (49), x(t) ∈ R6 is the state vector, u(t) ∈ R3 is the control input vector, Mij , i = 1 , ⋯ , 9 , j = 1 , ⋯ , 6 are the fuzzy sets defined as in Figures 1 and 2, Ai, i = 1 , ⋯ , 9 are obtained by sampling A(x(t)) in (46) at the given nine operating-points, and B is given in (47). With the normalized weight functions hi, i = 1 , ⋯ , 9 defined by

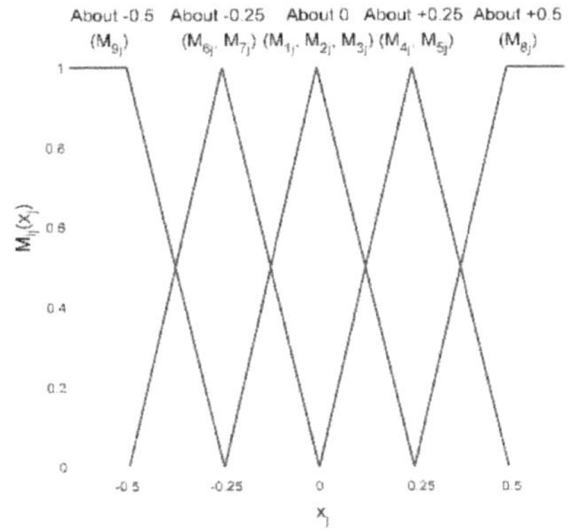

Figure 1. Fuzzy sets Mij, i = 1 , ⋯ , 9 , j = 1 , ⋯ , 3 used to design TS-type fuzzy model in (49) for system in (45).

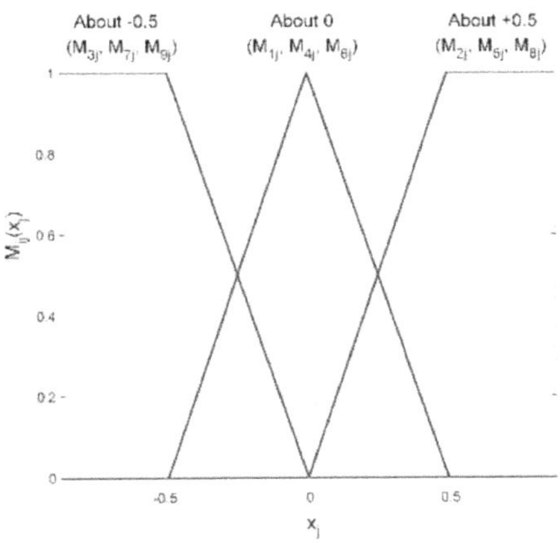

Figure 2. Fuzzy sets Mij , i = 1 , ⋯ , 9 , j = 4 , ⋯ , 6 used to design TS-type fuzzy model in (49) for system in (45).

$$h_i(x(t)) \triangleq \frac{\prod_{j=1}^{6} M_{ij}(x_j(t))}{\sum_{i=1}^{9} \prod_{j=1}^{6} M_{ij}(x_j(t))}, i = 1, \cdots, 9,$$

the TS-type fuzzy model in (49) can be transformed into the following TS-type fuzzy system:

$$\dot{x}(t) = \sum_{i=1}^{9} h_i(x(t)) A_i x(t) + Bu(t). \tag{49}$$

We can use the result of theorem 2 to design the optimal control law for the TS-type fuzzy system in (50) because the TS-type fuzzy system in (50) has the common input matrix B in (47). Then, we assume that = 2 , = 0.1 , and $R = I_{3\times3}$ in (28) and (29). With these values, the authors solve the LMIs in (27), (28), and (29) by using the command of "feasp" provided by the LMI Control Toolbox of MATLAB [35]. From the result, we obtain the following optimal control law for the TS-type fuzzy system in (43) that minimizes the cost function in (31):

$$u_a(t) = -K_a x(t), \tag{50}$$

where

$$K_a = \begin{bmatrix} 110.0629 & -24.5318 & -47.2469 & 73.0798 & -28.6127 & -37.7649 \\ -16.3545 & 139.9846 & -60.3201 & -17.0025 & 75.0011 & -39.4506 \\ -23.6235 & -45.2400 & 185.6703 & -5.6923 & -18.6642 & 92.8441 \end{bmatrix}$$

Since we can analyze the pure effect of the weight matrix R on the control performance without any constraint, the author designs two other optimal control laws for the TS-type fuzzy system in (50) without considering the decay rate constraint, which is given by (29), to clearly analyze the influence of the weight matrix R in (31) on the control performance. The author designs one by solving (27) and (28) for the TS-type fuzzy system in (50) with = 2 and $R = I_{3\times3}$, and one is given by

$$u_b(t) = -K_b x(t), \tag{51}$$

where

$$K_b = \begin{bmatrix} 28.0708 & 2.5099 & -1.9744 & 3.5899 & 0.3747 & -0.1420 \\ 1.6733 & 33.7936 & -4.9727 & 1.3456 & 1.7520 & 0.3518 \\ -0.9872 & -3.7295 & 42.8319 & 1.4713 & 1.3202 & -0.2517 \end{bmatrix}$$

The author designs the other by solving (27) and (28) for the TS-type fuzzy system in (50) with = 2 and $R = 3 \times I_{3\times3}$, and the other is given by

$$u_c(t) = -K_c x(t), \tag{52}$$

where

$$K_c = \begin{bmatrix} 29.1798 & 18.4270 & 18.9874 & 0.6854 & 0.2322 & -1.5053 \\ 12.2846 & 40.0169 & 18.1900 & 2.3547 & -1.7662 & -1.3049 \\ 9.4937 & 13.6425 & 49.3411 & 2.4969 & 0.9770 & -4.9648 \end{bmatrix}.$$

4.3. Numerical Simulation Results

As the numerical simulation model, the author uses the equations of rotational motion of a rigid body given by (38) and (41), where the inertia matrix is given by (48). The author assumes that the initial conditions at the initial time $t_0 = 0$ sec for the Euler axis $\hat{e}$ and the Euler angle ϕ are $\hat{e}(t_0) = [0.4896\ 0.2030\ 0.8480]^T$ and ϕ(t0) = 2.4648 rad, respectively, which give q(t0) = [0.3320 0.4618 0.1915 0.7999]T. Note that the given ϕ(t0) represents an almost upside-down initial orientation of a rigid body. Also, the author assumes a rest-to-rest maneuver of a rigid body, and, thus, the author assumes that the initial condition for the angular velocity vector is ω(t0) = [0 0 0]T rad/sec.

With the optimal control laws ua(t) of (51), ub(t) of (52), and uc(t) of (53), the author illustrates the influences of the decay rate in (29) and the weight matrix R in (31) on the control performance. Then, in Figures 3–5, the author shows the numerical simulation results for a rigid body with each control law ua(t), ub(t), and uc(t). In Figures 3–5, the red-solid, green-dashed, and blue-dotted lines represent the state trajectories of a rigid body with ua(t), ub(t), and uc(t), respectively.

First, as shown in Figures 3–5, we see that ua(t), ub(t), and uc(t) guarantee the asymptotic stability of the equilibrium point. Second, in Figures 3 and 4, we observe that ua(t) provides more desirable control performance than ub(t) and uc(t) because the design of ua(t) incorporates with the decay rate constraint.

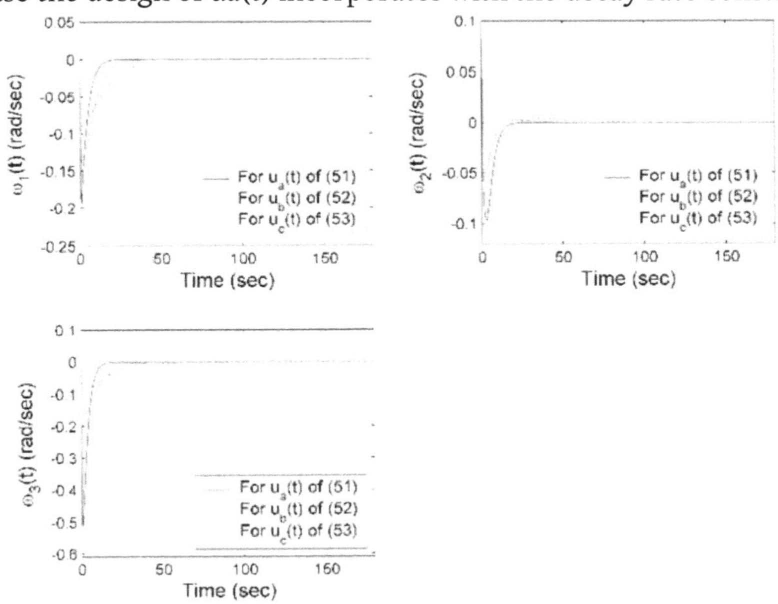

Figure 3.Angular velocity responses of the rigid body given by (38) and (41) with control laws ua(t) of (51), ub(t) of (52), and uc(t) of (53) designed in this chapter at initial conditions ω(t0) = [0 0 0]T rad/sec and q(t0) = [0.3320 0.4618 0.1915 0.7999]T .

Finally, in Figures 3 and 4, we see that the state responses of a rigid body with ub(t) show faster convergence rates to the equilibrium point than those with uc(t). And, in Figure 5, we see that the control efforts using ub(t) and uc(t) are comparable. The result can be explained by the weight matrix R, where small diagonal elements increase the penalty on the state vector and decrease the penalty on the control input vector, leading to short-term system stability. Conversely, a weight matrix R with large diagonal elements decreases the penalty on the state vector and increases the penalty on the control input vector, resulting in long-term system stability.

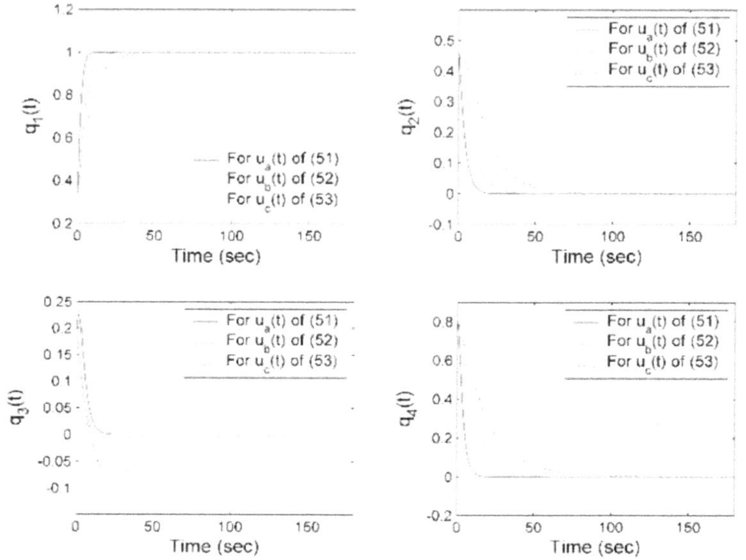

Figure 4. Quaternion responses of the rigid body given by (38) and (41) with control laws ua(t) of (51), ub(t) of (52), and uc(t) of (53) designed in this chapter at initial conditions ω(t0) = [0 0 0]T rad/sec and q(t0) = [0.3320 0.4618 0.1915 0.7999]T.

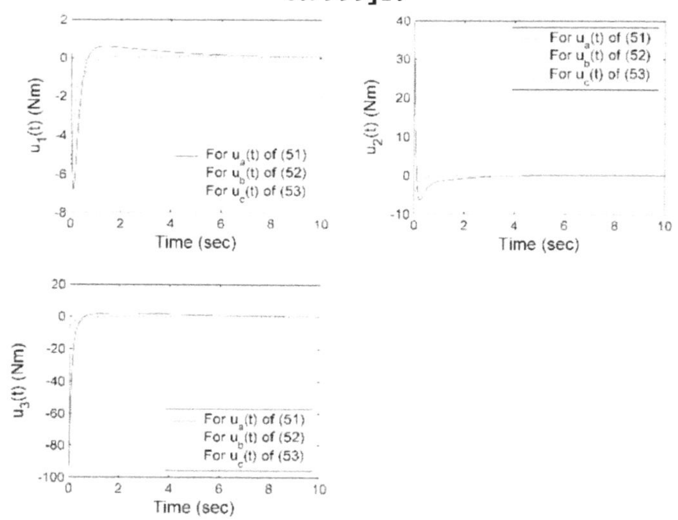

Figure 5.Control input responses of the rigid body given by (38) and (41) with control laws ua(t) of (51), ub(t) of (52), and uc(t) of (53) designed in this chapter at initial conditions ω(t0) = [0 0 0]T rad/sec and q(t0) = [0.3320 0.4618 0.1915 0.7999]T.

5. CONCISE SUMMARY

The author discussed a theory regarding the optimal control of nonlinear dynamic systems using dynamic programming and the inverse optimal technique in this chapter. The author used dynamic programming to formulate the Hamilton-Jacobi-Bellman (H-J-B) equation for the optimal control problem of nonlinear dynamic systems. They also adopted the inverse optimal approach to bypass the need for numerical solution of the H-J-B equation.

The author developed an optimal control architecture for TS-type fuzzy systems to provide global asymptotic stability of an equilibrium point, optimality in relation to a cost function, and fast convergence rates of state trajectories to an equilibrium point. The author introduced a systematic method for creating the optimal control law for TS-type fuzzy systems in this optimal control design. The author demonstrated the efficacy of optimal control design by examining the three-axis attitude stabilization issue of a stiff body. A control law was developed to stabilize the attitude of a rigid body along three axes, and its performance was evaluated by numerical simulations. The numerical simulation results showed that the three-axis attitude stabilizing control rule established in this chapter offers optimal control performance and fast convergence rates of state trajectories to an equilibrium point. The author proposes two more research topics: extending the studies in this chapter to develop resilient control designs for TS-type fuzzy systems considering system parametric uncertainties and external disruptions. The challenge is to integrate robustness considerations into the optimal control design methodology. The solution to this challenge can be explored in various literature sources that discuss robust control methods like loop-transfer-recovery, guaranteed-cost, stochastic, and state-estimation approaches. The challenge in integrating these approaches with the optimal control design method arises from their primary focus on linear systems. Hence, it may be necessary to apply similar methods to nonlinear dynamic systems. The other goal is to create the best control design for TS-type fuzzy systems to address tracking issues. Tracking issues arise when the equilibrium point is not at zero. In linear dynamic systems, tracking issues can be simplified to regulation problems if the intended state is known, or to disturbance-rejection problems if the disturbance signal is known. Often, the desired condition or disruption signal is unknown in many instances. In such situations, options like the minimax method and proportional-integral control may be necessary. The minimax strategy aims to design for the worst-case scenario by maximizing the disturbance signal that the control input minimizes, using the same performance metric. Proportional-integral control can be utilized to eliminate continual disruptions. In the following instalment, the minimax strategy and the proportional-integral control could offer a resolution to the tracking issue in TS-type fuzzy systems.

6. AT A GLANCE

This chapter discusses a theoretical finding on the optimal control of nonlinear dynamic systems. The author introduces the optimal control issue for nonlinear dynamic systems and demonstrates its solvability through dynamic programming and the inverse optimal technique. The author uses dynamic programming to establish the Hamilton-Jacobi-Bellman (H-J-B) equation related to the optimal control issue in nonlinear dynamic systems. The author then introduces an analytical method for solving the Hamilton-Jacobi-Bellman equation using the inverse optimal technique. The author uses a theoretical result in this chapter to develop an optimal control design for TS-type fuzzy systems. This design ensures global asymptotic stability of an equilibrium point, optimality in relation to a cost function, and fast convergence rates of state trajectories to the equilibrium point. The author examines the three-axis attitude stabilization problem of a stiff body to demonstrate the best control design approach for TS-type fuzzy systems. The author creates an ideal three-axis attitude stabilizing control rule for a stiff body using an optimal control design method and evaluates its control performance using numerical simulations.

REFERENCES

1. Zadeh LA. Fuzzy sets. Information and Control. 1965;8(3):338-353
2. Zadeh LA. Fuzzy algorithms. Information and Control. 1968;12(2):94-102
3. Bezdek JC. Editorial: Fuzzy models what are they, and why? IEEE Transactions on Fuzzy Systems. 1994;1(1):1-6
4. Khalil HK. Nonlinear Systems. Upper Saddle River, NJ: Prentice-Hall; 1996
5. Wang LX. Adaptive Fuzzy Systems and Control. Englewood Cliffs, NJ: Prentice-Hall; 1994
6. Zadeh LA. Outline of a new approach to the analysis of complex systems and decision processes. IEEE Transactions on Systems, Man, and Cybernetics. 1973;SMC-3(1):28-44
7. Mamdani EH, Assilian S. Applications of fuzzy algorithms for control of simple dynamic plant. Proceedings of the Institution of Electrical Engineers. 1974;121(12):1585-1588
8. Tanaka K, Sugeno M. Stability analysis and design of fuzzy control system. Fuzzy Sets and Systems. 1992;45(2):135-156
9. Wang H, Tanaka K, Griffin M. An approach to fuzzy control of nonlinear systems: Stability and design issues. IEEE Transactions on Fuzzy Systems. 1996;4(1):14-23

10. Feng G, Cao SG, Rees NW, Chak CK. Design of fuzzy control systems with guaranteed stability. Fuzzy Sets and Systems. 1997;85(1):1-10

11. Tanaka K, Ikeda T, Wang H. Fuzzy regulators and fuzzy observers: Relaxed stability conditions and LMI-based designs. IEEE Transactions on Fuzzy Systems. 1998;6(2):250-265

12. Joh J, Chen YH, Langari R. On the stability issues of linear Takagi-Sugeno fuzzy models. IEEE Transactions on Fuzzy Systems. 1998;6(3):402-410

13. Teixeira MCM, Żak SH. Stabilizing controller design for uncertain nonlinear systems using fuzzy models. IEEE Transactions on Fuzzy Systems. 1999;7(2):133-142

14. Park J, Kim J, Park D. LMI-based design of stabilizing fuzzy controllers for nonlinear systems described by Takagi-Sugeno fuzzy model. Fuzzy Sets and Systems. 2001;122(1):73-82

15. Takagi T, Sugeno M. Fuzzy identification of systems and its applications to modeling and control. IEEE Transactions on Systems, Man, and Cybernetics. 1985;SMC-15(1):116-132

16. Dorato P, Abdallah C, Cerone V. Linear-Quadratic Control: An Introduction. Upper Saddle River, NJ: Prentice-Hall; 1995

17. Wang LX. Stable and optimal fuzzy control of linear systems. IEEE Transactions on Fuzzy Systems. 1998;6(1):137-143

18. Wu SJ, Lin CT. Optimal fuzzy controller design: Local concept approach. IEEE Transactions on Fuzzy Systems. 2000;8(2):171-185

19. Wu SJ, Lin CT. Optimal fuzzy controller design: Global concept approach. IEEE Transactions on Fuzzy Systems. 2000;8(6):713-729

20. Park Y, Tahk MJ, Park J. Optimal stabilization of Takagi-Sugeno fuzzy systems with application to spacecraft control. Journal of Guidance, Control, and Dynamics. 2001;24(4):767-777

21. Kim D, Rhee S. Design of an optimal fuzzy logic controller using response surface methodology. IEEE Transactions on Fuzzy Systems. 2001;9(3):404-412

22. Wu SJ, Lin CT. Discrete-time optimal fuzzy controller design: Global concept approach. IEEE Transactions on Fuzzy Systems. 2002;10(1):21-38

23. Chen B, Liu X. Fuzzy guaranteed cost control for nonlinear systems with time-varying delay. IEEE Transactions on Fuzzy Systems. 2005;13(2):238-249

24. Mirzaei A, Moallem M, Mirzaeian B, Fahimi B. Design of an optimal fuzzy controller for antilock braking systems. IEEE Transactions on Vehicular Technology. 2006;55(6):1725-1730

25. Lin P-T, Wang C-H, Lee T-T. Time-optimal control of T-S fuzzy models via Lie algebra. IEEE Transactions on Fuzzy Systems. 2009;17(4):737-749

26. Mostefai L, Denai M, Oh S, Hori Y. Optimal control design for robust fuzzy friction compensation in a robot joint. IEEE Transactions on Industrial Electronics. 2009;56(10):3832-3839

27. Zhu Y. Fuzzy optimal control for multistage fuzzy systems. IEEE Transactions on Systems, Man, and Cybernetics, Part B: Cybernetics. 2011;41(4):964-975

28. Esfahani SH, Sichani AK. Improvement on the problem of optimal fuzzy Hi-tracking control design for non-linear systems. IET Control Theory and Applications. 2011;5(18):2179-2190

29. Bellman RE. The theory of dynamic programming. Proceedings of the National Academy of Sciences of the United States of America. 1952;38:716-719

30. Sepulchre R, Janković M, Kokotović PV. Constructive Nonlinear Control. New York: Springer-Verlag; 1997

31. Kalman RE. When is a linear control system optimal? Journal of Basic Engineering. 1964;86(1):51-60

32. Boyd S, Ghaoui El L, Feron E, Balakrishnan V. Linear Matrix Inequalities in System and Control Theory, SIAM Studies in Applied Mathematics Series. Vol. 15. Philadelphia, PA: SIAM; 1994

33. Vandenberghe L, Balakrishnan V. Algorithms and software for LMI problems in control. IEEE Control Systems Magazine. 1997;17(5):89-95

34. Nesterov Y, Nemirovskii A. Interior-Point Polynomial Algorithms in Convex Programming, SIAM Studies in Applied Mathematics Series. Vol. 13. Philadelphia, PA: SIAM; 1994

35. Gahinet P, Nemirovski A, Laub AJ, Chilali M. LMI Control Toolbox. Natick, MA: The MathWorks, Inc.; 1995

36. Junkins JL, Turner JD. Optimal Spacecraft Rotational Maneuvers, Studies in Astronautics. Vol. 3. Amsterdam, Netherlands: Elsevier Science; 1986

37. Shuster MD. A survey of attitude representations. Journal of Astronautical Sciences. 1993;41(4):439-517

38. Marandi SR, Modi V. A preferred coordinate system and the associated orientation representation in attitude dynamics. Acta Astronautica. 1987;15:833-843

39. Krstić M, Kanellakopoulos I, Kokotović PV. Nonlinear and Adaptive Control Design. New York: Wiley; 1995

40. Debs AS, Athans M. On the optimal angular velocity control of asymmetrical space vehicles. IEEE Transactions on Automatic Control. 1969;14(1):80-83

41. Dabbous TE, Ahmed NU. Nonlinear optimal feedback regulation of satellite angular momenta. IEEE Transactions on Aerospace and Electronic Systems. 1982;18(1):2-10

42. Vadali SR, Kraige LG, Junkins JL. New results on the optimal spacecraft attitude maneuver problem. Journal of Guidance, Control, and Dynamics. 1984;7(3):378-380

43. Athans M, Falb PL, Lacoss RT. Time-, fuel-, and energy-optimal control of nonlinear norm-invariant systems. IEEE Transactions on Automatic Control. 1963;8(3):196-202

44. Dixon MV, Edelbaum T, Potter JE, Vandervelde WE. Fuel optimal reorientation of axisymmetric spacecraft. Journal of Spacecraft and Rockets. 1970;7(11):1345-1351

45. Junkins JL, Carrington CK, Williams CE. Time-optimal magnetic attitude maneuvers. Journal of Guidance, Control, and Dynamics. 1981;4(4):363-368

46. Scrivener SL, Thomson RC. Survey of time-optimal attitude maneuvers. Journal of Guidance, Control, and Dynamics. 1994;17(2):225-233

47. Windeknecht TG. Optimal stabilization of rigid body attitude. Journal of Mathematical Analysis and Applications. 1963;6(2):325-335

48. Kumar KSP. On the optimum stabilization of a satellite. IEEE Transactions on Aerospace and Electronic Systems. 1965;1(2):82-83

49. Tsiotras P, Corless M, Rotea M. Optimal control of rigid body angular velocity with quadratic cost. In: Proceedings of the 35th IEEE Conference on Decision and Control; 11-13 December 1996; Kobe, Japan: pp. 1630-1635

50. Carrington CK, Junkins JL. Optimal nonlinear feedback control for spacecraft attitude maneuvers. Journal of Guidance, Control, and Dynamics. 1986;9(1):99-107

51. Rotea M, Tsiotras P, Corless M. Suboptimal control of rigid body motion with a quadratic cost. Dynamics and Control. 1998;8(1):55-81

52. Tsiotras P. Stabilization and optimality results for the attitude control problem. Journal of Guidance, Control, and Dynamics. 1996;19(4):772-779

53. Tsiotras P. Optimal regulation and passivity results for axisymmetric rigid bodies using two controls. Journal of Guidance, Control, and Dynamics. 1997;20(3):457-463

54. Bharadwaj S, Osipchuk M, Mease KD, Park FC. Geometry and inverse optimality in global attitude stabilization. Journal of Guidance, Control, and Dynamics. 1998;21(6):930-939

55. Park Y, Tahk MJ. Robust attitude stabilization of spacecraft using minimal kinematic parameters. In: Proceedings of the 2001 IEEE International Conference on Robotics and Automation; 21–26 May 2001; Seoul, Korea. pp. 1621-1626

56. Park Y, Tahk MJ. Optimal attitude stabilization of spacecraft using minimal kinematic parameters. In: Proceedings of the 4th Asian Control Conference; 25-27 September 2002; Singapore. pp. 881-885

57. Anderson BDO, Moore JB. Optimal Control: Linear Quadratic Methods. Englewood Cliffs, NJ: Prentice-Hall; 1990

58. Park Y, Tahk MJ, Bang HC. A game-theoretic approach to robust optimal attitude stabilization of a spacecraft with external disturbances. In: Proceedings of the JSASS 15th International Sessions in 39th Aircraft Symposium; 29–31 October 2001; Gifu, Japan. pp. 17-20

59. Wen JT, Kreutz-Delgado K. The attitude control problem. IEEE Transactions on Automatic Control. 1991;36(10):1148-1162

60. Lizarralde F, Wen JT. Attitude control without angular velocity measurement: A passivity approach. IEEE Transactions on Automatic Control. 1996;41(3):468-472

61. Caccavale F, Natale C, Siciliano B, Villani L. Six-DOF impedance control based on angle/axis representations. IEEE Transactions on Robotics and Automation. 1999;15(2):289-300

62. Park Y. Optimal control of TS-type fuzzy systems. IEEE Transactions on Aerospace and Electronic Systems. 2014;50(1):761-772

63. Horn R. Matrix Analysis. New York: Cambridge University Press; 1985

INDEX

www.ingramcontent.com/pod-product-compliance
Lightning Source LLC
LaVergne TN
LVHW062122200926
848484LV00007B/1743